PENNSYLVANIA COLLEGE OF TECHNOLOGY LIBRARY

5 0608 01133659 0

D1242478

ASE Test Preparation Series

Transit Bus Test

Electrical/Electronic Systems (Test H6)

THOMSON

DELMAR LEARNING

Australia   Canada   Mexico   Singapore   Spain   United Kingdom   United States

LIBRARY
Pennsylvania College
of Technology

One College Avenue
Williamsport, PA 17701-5799

SEP 2 0 2007

THOMSON

DELMAR LEARNING

Thomson Delmar Learning's ASE Test Preparation Series
Transit Bus Test for Electrical / Electronic Systems (Test H6)

Vice President, Technology
Professional Business Unit:
Gregory L. Clayton

Product Development Manager:
Kristen L. Davis

Product Manager:
Kimberley Blakey

Editorial Assistant:
Vanessa Carlson

Director of Marketing:
Beth A. Lutz

Marketing Manager:
Brian McGrath

Marketing Specialist:
Marissa Maiella

Marketing Coordinator:
Jennifer Stall

Production Director:
Patty Stephan

Production Manager:
Andrew Crouth

Content Project Manager:
Kara A. DiCaterino

Art Director:
Robert Plante

Cover Design
Michael Egan

COPYRIGHT © 2006 Thomson Delmar Learning. Thomson, the Star Logo, and Delmar Learning are trademarks used herein under license.

Printed in the United States of America
1 2 3 4 5 XX 10 09 08 07 06

For more information contact Thomson Delmar Learning
Executive Woods
5 Maxwell Drive, PO Box 8007,
Clifton Park, NY 12065-8007
Or find us on the World Wide Web at
www.delmarlearning.com

ALL RIGHTS RESERVED. No part of this work covered by the copyright hereon may be reproduced in any form or by any means—graphic, electronic, or mechanical, including photocopying, recording, taping, Web distribution, or information storage and retrieval systems—without the written permission of the publisher.

For permission to use material from the text or product, contact us by
Tel. (800) 730-2214
Fax (800) 730-2215
www.thomsonrights.com

Library of Congress Cataloging-in-Publication Data:
Card Number:

ISBN: 1-4180-4999-9

NOTICE TO THE READER

Publisher does not warrant or guarantee any of the products described herein or perform any independent analysis in connection with any of the product information contained herein. Publisher does not assume, and expressly disclaims, any obligation to obtain and include information other than that provided to it by the manufacturer.

The reader is expressly warned to consider and adopt all safety precautions that might be indicated by the activities herein and to avoid all potential hazards. By following the instructions contained herein, the reader willingly assumes all risks in connection with such instructions.

The publisher makes no representation or warranties of any kind, including but not limited to, the warranties of fitness for particular purpose or merchantability, nor are any such representations implied with respect to the material set forth herein, and the publisher takes no responsibility with respect to such material. The publisher shall not be liable for any special, consequential, or exemplary damages resulting, in whole or part, from the readers' use of, or reliance upon, this material.

Contents

Section 5 Sample Test for Practice

Section 6 Additional Test Questions for Practice

Section 7 Appendices

Preface

Delmar Learning is very pleased that you have chosen our ASE Test Preparation Guide to prepare yourself for the Transit Bus Brakes (H4) ASE Examination. This guide is designed to introduce you to the Task List for the Brakes (H4) test you are preparing to take, give you an understanding of what you are expected to be able to do in each task, and take you through sample test questions formatted in the same way the ASE tests are structured.

If you have a basic working knowledge of the discipline you are testing for, you will find Delmar Learning's ASE Test Preparation Guide to be an excellent way to understand the "must know" items to pass the test. This book is not a textbook. Its objective is to prepare the technician who has the requisite experience and schooling to challenge ASE testing. It cannot replace the hands-on experience or the theoretical knowledge required by ASE to master vehicle repair technology. If you are unable to understand more than a few of the questions and their explanations in this book, it could be that you require either more shop-floor experience or further study.

Each book begins with an item-by-item overview of the ASE Task List with explanations of the minimum knowledge you must possess to answer questions related to the task. Following that there are 2 sets of sample questions followed by an answer key to each test and an explanation of the answers to each question. A few of the questions are not strictly ASE format but were included because they help teach a critical concept that will appear on the test. We suggest that you read the complete Task List Overview before taking the first sample test. After taking the first test, score yourself and read the explanation to any questions that you were not sure about, including the questions you answered correctly. Each test question has a reference back to the related task or tasks that it covers. This will help you to go back and read over any area of the task list that you are having trouble with. Once you are satisfied that you have all of your questions answered from the first sample test, take the additional tests and check them. If you pass these tests, you will be prepared to do well on the ASE test.

Our Commitment to Excellence

Delmar Learning has sought out the best technicians in the country to help with the development of this 1st edition of the Transit Bus Brakes (H6) Test Preparation Guide.

Thank you for choosing Delmar Learning's ASE Test Preparation Guide. All of the writers, editors, and Delmar Staff have worked very hard to make this guide second to none. We know you are going to find this book accurate and easy to work with. It is our objective to constantly improve our product at Delmar by responding to feedback.

*If you have any questions concerning this book, please email us at:*autoexpert@trainingbay.com.

The History and Purpose of ASE

ASE began as the National Institute for Automotive Service Excellence (NIASE). It was founded as a non-profit independent entity in 1972 by a group of industry leaders with the single goal of providing a means for consumers to distinguish between incompetent and competent technicians. It accomplishes this goal by testing and certification of repair and service professionals. From this beginning it has evolved to be known simply as ASE (Automotive Service Excellence) and today offers more than 40 certification exams in automotive, medium/heavy duty truck, collision, engine machinist, school bus, parts specialist, automobile service consultant, and other industry-related areas. At this time there are more than 400,000 professionals with current ASE certifications. These professionals are employed by new car and truck dealerships, independent garages, fleets, service stations, franchised service facilities, and more. ASE continues its mission by also providing information that helps consumers identify repair facilities that employ certified professionals through its Blue Seal of Excellence Recognition Program. Shops that have a minimum of 75% of their repair technicians ASE certified and meet other criteria can apply for and receive the Blue Seal of Excellence Recognition from ASE.

ASE recognized that educational programs serving the service and repair industry also needed a way to be recognized as having the faculty, facilities, and equipment to provide a quality education to students wanting to become service professionals. Through the combined efforts of ASE, industry, and education leaders, the non-profit National Automotive Technicians Education Foundation (NATEF) was created to evaluate and recognize training programs. Today more than 2000 programs are ASE certified under the standards set by the service industry. ASE/NATEF also has a certification of industry (factory) training program known as CASE. CASE stands for Continuing Automotive Service Education and recognizes training provided by replacement parts manufacturers as well as vehicle manufacturers.

ASE certification testing is administered by the American College Testing (ACT). Strict standards of security and supervision at the test centers insure that the technician who holds the certification earned it. Additionally ASE certification also requires that the person passing the test to be able to demonstrate that they have two years of work experience in the field before they can be certified. Test questions are developed by industry experts that are actually working in the field being tested. There is more detail on how the test is developed and administered in the next section. Paper and pencil tests are administered twice a year at over seven hundred locations in the United States. Computer based testing is now also available with the benefit of instant test results at certain established test centers. The certification is valid for five years and can be recertified by retesting. So that consumers can recognize certified technicians, ASE issues a jacket patch, certificate, and wallet card to certified technicians and makes signs available to facilities that employ ASE certified technicians.

You can contact ASE at any of the following:

National Institute for Automotive Service Excellence
101 Blue Seal Drive S.E.
Suite 101
Leesburg, VA 20175
Telephone 703-669-6600
FAX 703-669-6123
www.ase.com

2 Take and Pass Every ASE Test

Participating in an Automotive Service Excellence (ASE) voluntary certification program gives you a chance to show your customers that you have the "know-how" needed to work on today's modern vehicles. The ASE certification tests allow you to compare your skills and knowledge to the automotive service industry's standards for each specialty area.

If you are the "average" automotive technician taking this test, you are in your mid-thirties and have not attended school for about fifteen years. That means you probably have not taken a test in many years. Some of you, on the other hand, have attended college or taken postsecondary education courses and may be more familiar with taking tests and with test-taking strategies. There is, however, a difference in the ASE test you are preparing to take and the educational tests you may be accustomed to.

How are the tests administered?

ASE test are administered at over 750 test sites in local communities. Paper and pencil tests are the type most widely available to technicians. Each tester is given a booklet containing questions with charts and diagrams where required. You can mark in this test booklet but no information entered in the booklet is scored. Answers are recorded on a separate answer sheet. You will enter your answers, using a number 2 pencil only. ASE recommends you bring four sharpened number 2 pencils that have erasers. Answer choices are recorded by coloring in the blocks on the answer sheet. The answer sheets are scanned electronically and the answers tabulated. For test security, test booklets include randomly generated questions. Your answer key must be matched to the proper booklet so it is important to correctly enter the booklet serial number on the answer sheet. All instructions are printed on the test materials and should be followed carefully.

ASE has introduced Computer Based Testing (CBT) at some locations. While the test content is the same for both testing methods the CBT tests have some unique requirements and advantages. It is strongly recommended that technicians considering the CBT tests go the ASE web page at www.ASE.com and review the conditions and requirements for this type of test. There is a demonstration of a CBT that allows you to experience this type of test before you register. Some technicians find this style of testing provides an advantage, while others find operating the computer a distraction. One significant benefit of CBT is the availability of instant results. You can receive your test results before you leave the test center. CBT testing also offers increased flexibility in scheduling. The cost for taking CBTs is slightly higher than paper and pencil tests and the number of testing sites is limited. The first time test taker may be more comfortable with the paper and pencil tests but technicians now have a choice.

Who Writes the Questions?

The questions are written by service industry experts in the area being tested. Each area will have its own technical experts. Questions are entirely job related. They are designed to test the skills you need to be a successful technician. Theoretical knowledge is important and necessary to answer the questions, but the ability to apply that knowledge is the basis of ASE test questions.

Each question has its roots in an ASE "item-writing" workshop where service representatives from automobile manufacturers (domestic and import), aftermarket parts and equipment manufacturers, working technicians, and vocational educators meet in a workshop setting to share ideas and translate them into test questions. Each test question written by these experts must survive review by all members of the group.

The questions are written to deal with practical application of soft skills and system knowledge experienced by technicians in their day-to-day work.

All questions are pre-tested and quality-checked on a national sample of technicians. Those questions that meet ASE standards of quality and accuracy are included in the scored sections of the tests; the "rejects" are sent back to the drawing board or discarded altogether.

Each certification test is made up of between forty and eighty multiple-choice questions.

Note: Each test could contain additional questions that are included for statistical research purposes only. Your answers to these questions will not affect your score, but since you do not know which ones they are, you should answer all questions on the test. The five-year Recertification Test will cover the same content areas as those listed above. However, the number of questions in each content area of the Recertification Test will be reduced by about one-half.

Objective Tests

A test is called an objective test if the same standards and conditions apply to everyone taking the test and there is only one correct answer to each question.

Objective tests primarily measure your ability to recall information. A well-designed objective test can also test your ability to understand, analyze, interpret, and apply your knowledge. Objective tests include true-false, multiple choice, fill in the blank, and matching questions. ASE's tests consist exclusively of four-part multiple-choice objective questions.

The following are some strategies that may be applied to your tests.

Before beginning to take an objective test, quickly look over the test to determine the number of questions, but do not try to read through all of the questions. In an ASE test, there are usually between forty and eighty questions, depending on the subject. Read through each question before marking your answer. Answer the questions in the order they appear on the test. Leave the questions blank that you are not sure of and move on to the next question. You can return to those unanswered questions after you have finished the others. They may be easier to answer at a later time after your mind has had additional time to consider them on a subconscious level. In addition, you might find information in other questions that will help you recall the answers to some of them.

Do not be obsessed by the apparent pattern of responses. For example, do not be influenced by a pattern like **D, C, B, A, D, C, B, A** on an ASE test.

There is also a lot of folk wisdom about taking objective tests. For example, there are those who would advise you to avoid response options that use certain words such as *all, none, always, never, must,* and *only,* to name a few. This, they claim, is because nothing in life is exclusive. They would advise you to choose response options that use words that allow for some exception, such as *sometimes, frequently, rarely, often, usually, seldom,* and *normally.* They would also advise you to avoid the first and last option (A and D) because test writers, they feel, are more comfortable if they put the correct answer in the middle (B and C) of the choices. Another recommendation often offered is to select the option that is either shorter or longer than the other three choices because it is more likely to be correct. Some would advise you to never change an answer since your first intuition is usually correct.

Although there may be a grain of truth in this folk wisdom, ASE test writers try to avoid them and so should you. There are just as many **A** answers as there are **B** answers, just as many **D** answers as **C** answers. As a matter of fact, ASE tries to balance the answers at about 25 percent per choice **A, B, C,** and **D.** There is no intention to use "tricky" words, such as outlined above. Put no credence in the opposing words "sometimes" and "never," for example.

Multiple-choice tests are sometimes challenging because there are often several choices that may seem possible, and it may be difficult to decide on the correct choice. The best strategy, in this case, is to first determine the correct answer before looking at the options. If you see the answer you decided on, you should still examine the options to make sure that none seem more correct than yours. If you do not know or are not sure of the answer, read each option very carefully and try to eliminate those options that you know to be wrong. That way, you can often arrive at the correct choice through a process of elimination.

If you have gone through all of the test and you still do not know the answer to some of the questions, <u>then guess.</u> Yes, guess. You then have at least a 25 percent chance of being correct. If you leave the question blank, you have no chance. Your score is based on the number of questions answered correctly.

Preparing for the Exam

The main reason we have included so many sample and practice questions in this guide is, simply, to help you learn what you know and what you don't know. We recommend that you work your way through each question in this book. Before doing this, carefully look through Section 3; it contains a description and explanation of the question types you'll find on an ASE exam.

Once you understand what the questions will look like, move to the sample test. Answer one of the sample questions (Section 5) then read the explanation (Section 7) to the answer for that question. If you don't feel you understand the reasoning for the correct answer, go back and read the overview (Section 4) for the task that is related to that question. If you still don't feel you have a solid understanding of the material, identify a good source of information on the topic, such as a textbook, and do some more studying.

After you have completed all of the sample test items and reviewed your answers, move to the additional questions (Section 6). This time answer the questions as if you were taking an actual test. Do not use any reference or allow any interruptions in order to get a feel for how you will do on an actual test. Once you have answered all of the questions, grade your results using the answer key in Section 7. For every question that you gave a wrong answer to, study the explanations to the answers and/or the overview of the related task areas. Try to determine the root cause for your missing the question. The easiest thing to correct is learning the correct technical content. The hardest thing to correct is behaviors that lead you to a wrong conclusion. If you knew the information but still got it wrong there is a behavior problem that will need to be corrected. An example would be reading too quickly and skipping over words that affect your reasoning. If you can identify what you did that caused you to answer the question incorrectly you can eliminate that cause and improve your score. Here are some basic guidelines to follow while preparing for the exam:

- Focus your studies on those areas you are weak in.

- Be honest with yourself while determining if you understand something.

- Study often but in short periods of time.

- Remove yourself from all distractions while studying.

- Keep in mind the goal of studying is not just to pass the exam, the real goal is to learn!

- Prepare physically by getting a good night's rest before the test and eat meals that provide energy but do not cause discomfort.

- Arrive early to the test site to avoid long waits as test candidates check in and to allow all of the time available for your tests.

During the Test

On paper and pencil tests you will be placing your answers on a sheet where you will be required to color in your answer choice. Stray marks or incomplete erasures may be picked up as an answer by the electronic reader, so be sure only your answers end up on the sheet. One of the biggest problems an adult faces in test taking, it seems, is placing the answer in the correct spot on the answer sheet. Make certain that you mark your answer for, say, question 21, in the space on the answer sheet designated for the answer for question 21. A correct response in the wrong line will probably result in two questions being marked wrong, one with two answers (which could include a correct answer but will be scored wrong) and the other with no answer. Remember, the answer sheet on the written test is machine scored and can only "read" what you have colored in.

If you finish answering all of the questions on a test and have remaining time, go back and review the answers to those questions that you were not sure of. You can often catch careless errors by using the remaining time to review your answers. Carefully check your answer sheet for blank answer blocks or missing information.

At practically every test, some technicians will invariably finish ahead of time and turn their papers in long before the final call. Some technicians may be doing recertification tests and others may be taking fewer tests than you. Do not let them distract or intimidate you.

It is not wise to use less than the total amount of time that you are allotted for a test. If there are any doubts, take the time for review. Any product can usually be made better with some additional effort. A test is no exception. It is not necessary to turn in your test paper until you are told to do so.

Your Test Results!

You can gain a better perspective about tests if you know and understand how they are scored. ASE's tests are scored by American College Testing (ACT), a nonpartial, unbiased organization having no vested interest in ASE or in the automotive industry.

Each question carries the same weight as any other question. For example, if there are fifty questions, each is worth 2 percent of the total score. The passing grade is 70 percent. That means you must correctly answer thirty-five of the fifty questions to pass the test.

The test results can tell you:

- where your knowledge equals or exceeds that needed for competent performance, or

- where you might need more preparation.

Your ASE test score report is divided into content areas and will show the number of questions in each content area and how many of your answers were correct. These numbers provide information about your performance in each area of the test. However, because there may be a different number of questions in each content area of the test, a high percentage of correct answers in an area with few questions may not offset a low percentage in an area with many questions.

It should be noted that one does not "fail" an ASE test. The technician who does not pass is simply told "More Preparation Needed." Though large differences in percentages may indicate problem areas, it is important to consider how many questions were asked in each area. Since each test evaluates all phases of the work involved in a service specialty, you should be prepared in each area. A low score in one area could keep you from passing an entire test.

There is no such thing as average. You cannot determine your overall test score by adding the percentages given for each task area and dividing by the number of areas. It doesn't work that way because there generally are not the same number of questions in each task area. A task area with twenty questions, for example, counts more toward your total score than a task area with ten questions.

Your test report should give you a good picture of your results and a better understanding of your strengths and weaknesses for each task area.

If you fail to pass the test, you may take it again at any time it is scheduled to be administered. You are the only one who will receive your test score. Test scores will not be given over the telephone by ASE nor will they be released to anyone without your written permission.

3 Types of Questions on an ASE Exam

ASE certification tests are often thought of as being tricky. They may seem to be tricky if you do not completely understand what is being asked. The following examples will help you recognize certain types of ASE questions and avoid common errors.

Paper-and-pencil tests and computer-based test questions are identical in content and difficulty. Most initial certification tests are made up of forty to eighty multiple-choice questions. Multiple-choice questions are an efficient way to test knowledge. To answer them correctly, you must think about each choice as a possibility, and then choose the one that best answers the question. To do this, read each word of the question carefully. Do not assume you know what the question is about until you have finished reading it.

About 10 percent of the questions on an actual ASE exam will use an illustration. These drawings contain the information needed to correctly answer the question. The illustration must be studied carefully before attempting to answer the question. Often, technicians look at the possible answers then try to match up the answers with the drawing. Always do the opposite; match the drawing to the answers. When the illustration is showing an electrical schematic or another system in detail, look over the system and try to figure out how the system works before you look at the question and the possible answers.

Multiple-Choice Questions

The most common type of question used on ASE Tests is the multiple-choice question. This type of question contains three "distracters" (wrong answers) and one "key" (correct answer). When the questions are written effort is made to make the distracters plausible to draw an inexperienced technician to one of them. This type of question gives a clear indication of the technician's knowledge. Using multiple criteria including cross-sections by age, race, and other background information, ASE is able to guarantee that a question does not bias for or against any particular group. A question that shows bias toward any particular group is discarded. If you encounter a question that you are unsure of, reverse engineer it by eliminating the items that it cannot be. For example:

When considering a bus service brake relay valve:

A. air lines are connected from the service brake relay valve delivery ports to the rear axle service brake chambers.
B. when the brakes are released the inlet valve is open in the service brake relay valve.
C. when the brakes are released the exhaust valve is closed in the service brake relay valve.
D. the service brake relay valve is in a balanced position when the air pressure in the rear brake chambers equals reservoir pressure.

Answer:

Answer A is correct. Air lines are connected from the service brake relay valve delivery ports to the rear axle service brake chamber.
Answer B is wrong. When the brakes are released the inlet valve is closed, not open, in the service brake relay valve.

Answer C is wrong. When the brakes are released the exhausts valve is open not closed in the service brake relay valve.

Answer D is wrong. The service brake relay valve is in a balanced position when the air pressure in the rear brake chambers equals brake application pressure, not system pressure.

EXCEPT Questions

Another type of question used on ASE tests has answers that are all correct except one. The correct answer for this type of question is the answer that is wrong. The word "**EXCEPT**" will always be in capital letters. You must identify which of the choices is the wrong answer. If you read quickly through the question, you may overlook what the question is asking and answer the question with the first correct statement. This will make your answer wrong. An example of this type of question and the analysis is as follows:

When diagnosing a service brake relay valve, all of the following apply **EXCEPT**:

A. inlet valve leakage is tested with the service brakes released.
B. apply a soap solution to the area around the inlet and exhaust valve retaining ring to check exhaust valve leakage.
C. exhaust valve leakage is tested with the brakes applied.
D. the control port on the service brake relay valve is connected to the supply port on the brake application valve.

Answer:

Answer A is wrong. The inlet valve leakage is tested with the service brakes released.
Answer B is wrong. You do apply a soap solution to the area around the inlet and exhaust valve retaining ring to check exhaust valve leakage.
Answer C is wrong. Exhaust valve leakage is tested with the brakes applied.
Answer D is correct. This is the exception because the control port on the service brake relay valve is connected to the delivery port on the brake application valve.

Technician A, Technician B Questions

The type of question that is most popularly associated with an ASE test is the "Technician A says . . . Technician B says . . . Who is right?" type. In this type of question, you must identify the correct statement or statements. To answer this type of question correctly, you must carefully read each technician's statement and judge it on its own merit to determine if the statement is true.

Sometimes this type of question begins with a statement about some analysis or repair procedure. This is often referred to as the stem of the question and provides the setup or background information required to understand the conditions the question is based on. This is followed by two statements about the cause of the concern, proper inspection, identification, or repair choices. You are asked whether the first statement, the second statement, both statements, or neither statement is correct. Analyzing this type of question is a little easier than the other types because there are only two ideas to consider although there are still four choices for an answer.

Technician A, Technician B questions are really double true or false questions. The best way to analyze this kind of question is to consider each technician's statement separately. Ask yourself, is A true or false? Is B true or false? Then select your answer from the four choices. An important point to remember is that an ASE Technician A, Technician B question will never have Technician A and B directly disagreeing with each other. That is why you must evaluate each statement independently.

An example of this type of question and the analysis of it follows.

While discussing single and double check valves, Technician A says single check valves are connected between the supply and the primary and secondary reservoirs. Technician B says a double check valve allows air pressure to flow from the lowest of two pressure sources. Who is correct?

A. A only
B. B only
C. Both A and B
D. Neither A nor B

Answer:

Answer A is correct. Single check valves are connected between the supply and secondary reservoirs.
Answer B is wrong. A double check valve allows air pressure to flow from one of two sources, selecting the higher of the two and routing it to its outlet.
Answer C is wrong. Only Technician A is correct.
Answer D is wrong. Only Technician A is correct.

Most-Likely Questions

Most-Likely questions are somewhat difficult because only one choice is correct while the other three choices are nearly correct. An example of a Most-Likely-cause question is as follows:

During a brake inspection, a technician tests the air supply system and finds that the buildup time is slow. Which of the following is the Most-Likely cause?

A. A clogged compressor inlet filter
B. A leak in a brake chamber
C. An air leak in the secondary delivery line
D. A restriction in the governor line

Answer:

Answer A is correct. A clogged compressor inlet filter is the Most-Likely cause of slow buildup because it stops the compressor cylinder from breathing efficiently.
Answer B is wrong. The brake chamber is not part of the air supply system and cannot cause a slow buildup.
Answer C is wrong. The secondary delivery line is not part of the air supply circuit and cannot cause a slow buildup.
AnswerD is wrong. A restriction in the governor line could more likely cause high system pressure.

LEAST-Likely Questions

Notice that in Most-Likely questions there is no capitalization. This is not so with LEAST-Likely type questions. For this type of question, look for the choice that would be the LEAST-Likely cause of the described situation. Read the entire question carefully before choosing your answer. An example is as follows:

Which of the following is the LEAST-Likely cause of wheel bearing failure?

A. Overloading
B. Contamination
C. A damaged race/axle housing
D. Improper lubricant

Answer:

Answer A is correct. Overloading is the LEAST-Likely cause of a wheel bearing failure because they are designed with a significant margin of safety.
Answer B is wrong. Contamination will cause high heat buildup, destroying the bearing.
Answer C is wrong. Axle housing damage may cause wheel bearings to fail.
Answer D is wrong. Improper or incompatible lubricant will cause failure of the bearing.

Summary

There are no four-part multiple-choice ASE questions having "none of the above" or "all of the above" choices. ASE does not use other types of questions, such as fill-in-the-blank, completion, true-false, word-matching, or essay. ASE does not require you to draw diagrams or sketches. If a formula or chart is required to answer a question, it is provided for you. There are no ASE questions that require you to use a pocket calculator.

Testing Time Length

An ASE written test session is four hours. You may attempt from one to a maximum of four tests in one session. It is recommended, however, that no more than a total of 225 questions be attempted at any test session. This will allow for just over one minute for each question.

Visitors are not permitted at any time. If you wish to leave the test room, for any reason, you must first ask permission. If you finish your test early and wish to leave, you are permitted to do so only during specified dismissal periods.

You should monitor your progress and set an arbitrary limit to how much time you will need for each question. This should be based on the number of questions you are attempting. It is suggested that you wear a watch because some facilities may not have a clock visible to all areas of the room.

Computer-Based Tests are allotted a testing time according to the number of questions ranging from one half hour to one and one half hours. Advanced level tests are allowed two hours. This time is by appointment and you should be sure to be on time to ensure that you have all of the time allocated. If you arrive late for a CBT test appointment you will only have the amount of time remaining on your appointment.

4 Overview of Task List

Electrical and Electronic Systems (Test H6)

The following section includes the task areas and task lists for this test and a written overview of the topics covered in the test.

The task list describes the actual work you should be able to do as a technician that you will be tested on by the ASE. This is your key to the test and you should review this section carefully. We have based our sample test and additional questions upon these tasks. The overview section will also support your understanding of the task list. ASE advises that the questions on the test may not equal the number of tasks listed; the task lists tell you what ASE expects you to know how to do and be ready to be tested on.

At the end of each question in the Sample Test and Additional Test Questions sections, a letter and number will be used as a reference back to this section for additional study. Note the following example: **A2.**

A. General Electrical Diagnosis (20 Questions)

Task A2   **Check continuity in electrical/electronic circuits using appropriate test equipment.**

Example
1. Which of the following is MOST likely to be done when making a quick check of circuit continuity?
   A. connect an ammeter in parallel
   B. connect a voltmeter in series
   C. turn the ignition on
   D. use an ohmmeter   (A.2)

Analysis:

Question #1
Answer A is wrong. An ammeter is connected in series.
Answer B is wrong. Voltmeters are connected across the circuit for voltage or voltage drop testing.
Answer C is wrong. The ignition does not necessarily have to be on for a continuity check unless the circuit being tested is in series with the ignition switch.
Answer D is correct. An ohmmeter is used to check continuity.

Task List and Overview

A. General Electrical Diagnosis (20 Questions)

Task A1   **Verify operator complaint, reproduce the condition (including intermittent problems), and/or road test vehicle; determine necessary action.**

Bus operators are required by law to conduct pre-trip inspections of their vehicles on a daily basis. Transit agencies typically provide a written checklist for operators to guide them through the required inspection procedures and to note any deficiencies. When finished driving a particular vehicle, operators sign the completed checklists (sometimes referred to as defect cards) and return them to the appropriate agency location. Any noted deficiencies that would compromise safety, however, must be reported immediately before the vehicle enters service; other non-safety-related defects are reviewed by the maintenance department and scheduled for repair. If uncertain about the conditions surrounding the problem, technicians should obtain more information from that operator, especially if the problem is an intermittent one. Doing so helps to create a team approach to keeping buses mechanically sound and in safe operating condition.

Items included for daily vehicle pre- and post-trip inspections by the operator include an examination of tires, lights, mirrors, windshields and windows, fluid leaks, mobility aid/wheelchair/lift/ramp and securement equipment, safety equipment (signage, seat belts, fire extinguishers, etc.), brake system, and steps and handrails, among others.

Task A2   **Check continuity in electrical/electronic circuits using appropriate test equipment.**

Checking continuity in an electrical circuit is one of the most common troubleshooting tests a technician will make. While the purpose of this test is to make sure that a complete current path exists in the circuit being tested, it is important to remember that it is not an accurate indication of circuit performance (e.g., excessive resistance). A continuity test is most useful to quickly differentiate one circuit from another, such as trying to locate a specific contact in a multiple-pin connector.

Continuity tests can be made a number of different ways. If the circuit is energized, a 12/24-volt test light or voltmeter can be used to check for voltage at various test points. If the circuit is not energized and disconnected at both ends (the most common way to check continuity), then a self-powered test light or ohmmeter can be used to check for a complete circuit. Both tools accomplish this by energizing the circuit with low potential current to determine that the circuit can be closed. Many DMMs (digital multimeters) have a separate feature on them that will allow continuity tests to be made simply by listening for an audible beep. This is handy because multiple tests can be made rather quickly without having to look constantly at the display for a resistance value.

When testing circuits that include ECMs (electronic control modules), it is essential to observe the OEM diagnostic procedure to avoid possible damage to the processor.

Task A3   **Check applied voltages, circuit voltages, and voltage drops in electrical/electronic circuits using a digital multimeter (DMM).**

Voltage tests are measured with the test leads of the DMM in parallel to the component or circuit being tested. In testing applied voltage, the negative or black test lead of the DMM is connected to a battery or chassis ground. The positive red lead is then probed near the power source (e.g., a switch or fuse) to determine if the circuit is receiving the proper voltage. A handy feature of a DMM is that polarity is not important. If the leads are reversed, the display will simply include a minus sign in front of the reading. Circuit voltages are tested much the same way. Take the positive test probe and go from one circuit's power source to the next. Generally, the applied voltages should all be the same.

When selecting a DMM to use for testing electronic circuits, it is recommended that it have a 10 megohm or higher impedance. This is necessary to limit the affect the unit might have on accuracy when testing low current flow circuits. For this reason, never use a test light to check for voltage in an electronic circuit. The lamp in the test light will draw too much current and this can damage the circuit integrity.

If you need to test for voltage in an energized circuit that is in operation, a convenient way to do this is with a back probe tool, also known as a "spoon," at various connectors in a particular circuit. This DMM accessory allows you to probe into the connector from behind without disconnecting it.

Voltage drops are often misunderstood. A voltage drop is simply the loss of voltage (electrical "pressure") in a circuit due to resistance. Remember, that for there to be a voltage drop, there has to be current flow through the circuit being tested. Any resistance other than the designated load (e.g., light bulb, blower motor) will produce unwanted voltage drop along with reduced current flow in that circuit. As with the voltage test, the test leads are probed in parallel to the component or portion of the circuit being tested. If, for example, two mating contacts in a joined connector assembly were suspect, then both sides of the connector would be probed with the test leads. Any voltage reading with the circuit under load is a voltage drop. As a rule of thumb, each connection, switch, fuse, length of wire, etc., should measure no more than one-tenth of a volt (100 millivolts) drop while under load.

The most common diagnostic tool that every technician should have is a digital multimeter, which combines many test functions. An ammeter measures current, a voltmeter measures the potential difference (voltage) between two points, and an ohmmeter measures resistance. A multimeter combines these functions into a single instrument. Unlike an old fashioned analog meter with a needle that moves across a scale, a digital multimeter (DMM) displays an output in numbers, usually on a liquid crystal display.

Before the introduction of electronic circuits, the test light was a popular tool to diagnose for breaks in wires or loss of contact within a basic electrical circuit. However, the brightness of the bulb in the test light is not a measurement. Also, the test light itself needs approximately 1 amp to light its bulb, therefore making it useless on lower amperage electronic circuits. WARNING: Since digital multimeters are now readily available, there should be very limited areas where a technician needs to use a test light. In fact, some agencies do not allow technicians to have a test lamp in their tool box. In any case, never use test lights on computer circuits because damage to components may occur. The test light can act like a jumper wire connected directly to ground, causing excess current. If a test light is used on more traditional electrical circuits (i.e., those that do not involve electronic modules), be careful not to puncture or penetrate wiring if you can avoid it. This allows moisture to enter the insulation, causing corrosion and eventually a failure.

Task A4   Check current flow in electrical/electronic circuits and components using an ammeter, digital multimeter (DMM), or clamp-on ammeter.

Current flow tests are used when a circuit is suspected of having higher-than-normal current flow, such as a dragging blower motor or a circuit with a low resistance short to ground, in other words, situations that might continually blow a fuse or trip a circuit breaker.

In testing current flow with an ammeter or DMM, it is important to remember that the test leads are connected in series with the circuit being tested, usually at a point near the power source. The circuit must be interrupted at some point to allow the connection of the test leads.

Most DMMs have a 10- to 20-amp limit when measuring amperage directly through the meter. Any greater current flow will blow the DMM's fuse. If you suspect that the circuit carries more than that, then a safe way to test for current flow would be to use a current clamp. This device simply clamps over the wire being tested and determines current flow by measuring the strength of the magnetic field surrounding the wire. While it is extremely handy to use, it is not as accurate as routing all the current through the DMM, especially in circuits flowing less than 10 amps.

Remember: Low current flow usually is a result of excessive resistance in a circuit or low voltage. Higher-than-normal current flow can generally be traced to excessive applied voltage or a shorted component or wire.

Task A5   Check electronic circuit waveforms using an oscilloscope or graphing multimeter (GMM); interpret readings and determine needed repairs.

An oscilloscope or graphing multimeter (GMM) is a device used to produce visible patterns that are the graphical representations of electrical signals or waveforms. The graphs plot the relationships between two variables, usually voltage as a function of time. The vertical axis (scale) usually defines voltage generated by the input signal to the oscilloscope and the horizontal axis (scale) normally defines time. In its simplest mode, the oscilloscope repeatedly draws a horizontal line called a trace across the screen from left to right. The time it takes each trace to go across the screen is called a sweep. The most common types of signal waveforms are sinusoidal or sine waves (alternating current) and square waves (digital communications). Distortion of a signal can be caused by interference (electrical noise) as well as defective components.

An oscilloscope may contain a cathode-ray tube (CRT) or a digital liquid crystal display (LCD) screen. The graphing multimeter (GMM) is a digital handheld device (sometimes called a scope-meter) that contains a digital LCD screen. The GMM also doubles as a normal multimeter. Speed of response is the cathode-ray oscilloscope's main advantage over other graphing devices. Most oscilloscopes have plotting frequencies between 50 and 100 megahertz (MHz) on average. The screen displays a grid with lines and individual marks called graticles. The full-length vertical and horizontal graticles are called divisions. Each graticle represents two-tenths or 0.2 of the volts/division on the vertical scale and seconds/division on the horizontal scale.

Oscilloscopes will have two or more separate signal inputs called channels. It is sometimes necessary or desirable to display more than one signal at the same time on the screen of an oscilloscope. Using a variety of techniques, two or more signals can be simultaneously shown. This is especially useful in comparing a known good signal versus a suspected bad signal.

A scope probe is used to measure the signal. Scope probes can measure different voltage ranges and will be marked with an X1, X10, or have a switch to select the range. This only affects the voltage scale and not the time scale. The oscilloscope must be set up correctly to display the correct voltage scale or you must remember when visually measuring the scale. Since each probe has its own electrical characteristics (that can change over time), it is wise to often calibrate each probe to its own input channel that's being used. If not, the signal will be displayed inaccurately. Most scope probes have a ground clip attached near the tip. These ground clips are internally connected and common to each other and to the oscilloscope's chassis. Damage to the circuit may result if care is not taken to isolate these grounds. This is especially true when measuring balanced digital data communication lines.

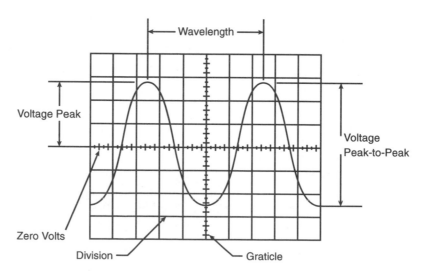

Oscilloscopes have adjustments for the voltage and time scales measured in volts/div and sec/div. Another adjustment is the trigger, which synchronizes the signals to a reference. These are adjusted to display the desired signal. Most digital oscilloscopes and GMMs will have an Auto-Set button. This will display the best possible signal of several wavelengths. One wavelength is the top or peak of a signal to the peak of the next and measured by time in seconds. The frequency (measured in Hertz) of a signal is inversely proportional to time (i.e. frequency = 1/time). The overall voltage of the signal is measured from the top of the signal to the bottom of the signal. This is called the peak-to-peak voltage. Voltage measured from the ground or zero-volt reference to either one of the top or the bottom of the signal is called the peak voltage. You may measure the waveform by counting the divisions and graticles in reference to the volts/div and sec/div. Some oscilloscopes and GMMs also display a numerical digital readout.

Task A6   Check resistance in electrical/electronic circuits and components using an ohmmeter or a digital multimeter (DMM).

Resistance checks are typically made when a circuit has unwanted voltage drops or low current flow. An ohmmeter is a device that circulates a small current through a circuit when it is not energized and then measures the voltage drop through it. It displays this resistance (or restriction to current flow) in units known as ohms. The lower the ohmic value, the less restriction to electron flow there will be in a circuit. If the meter reads infinity (or a high flashing number on most DMMs), this means that the circuit is open. Except for where a resistance is built into a circuit, such as a blower motor resistor or a load itself, generally speaking the lower resistance in a circuit, the better. For example, when testing a length of wire or a fuse, most will test very near zero ohms. When testing with a DMM, always be sure to zero the meter first to compensate for any resistance present in the test leads, especially when testing low resistance components or circuits.

When making tests with a DMM that is not auto-ranging, be sure to select a range that will provide accuracy. If you are using an analog meter, set the meter to a range setting that will put the needle in roughly the middle of the scale for the component you are testing.

Resistance checking of specific components is generally used when a manufacturer specifies a certain test value, such as a fuel level sending unit. Some components, such as light bulbs and glow plugs, do not lend themselves to resistance testing because their resistance changes as they heat up. Also, large diameter conductors such as battery cables cannot be reliably tested with an ohmmeter because it cannot circulate enough current to simulate actual operating conditions and identify resistance. Testing of battery cables is best done using a voltage drop test (see Task C1).

Task A7   Locate shorts, grounds, and opens in electrical/electronic circuits.

A short circuit is defined as one where the current flow is allowed to ground at a point other than where it was designed, such as a bare wire rubbing against the frame. An open circuit is usually caused by a broken wire or other component not making the necessary connection to complete a circuit, stopping current flow in that portion of the circuit.

Finding shorts is best done with an ohmmeter because in a live circuit, a fuse or breaker will continually blow. Do not install a larger fuse because there will be a risk of melting a bundle of wires. Divide the circuit into small sections at various connectors (where applicable) while testing for continuity to ground (there should be none), or visually inspect the harness for rub or pinch points.

Locating opens can sometimes be more difficult, because sometimes the damage is not visually apparent. A good way to test for an open would be to apply voltage at one end and then probe at each succeeding connection downstream until you find no reading. Alternatively, an ohmmeter can be used to do the same thing when the circuit is not energized.

It is important to remember when testing at connectors, especially smaller ones designed for electronic circuits, that you do not damage the contacts when probing into them, especially female ones. Spreading the tangs on a female contact while testing can create problems. Always use an appropriate adapter for testing these types of contacts.

Task A8   Diagnose battery drain problems with the master/key switch off.

While dead batteries are not always caused by a master/key switch problem, when you suspect this it is a good idea before you start troubleshooting to know the possible causes and how the system is constructed. Before the widespread use of electronics in vehicles, many systems had zero current draw with the master/key switch off. Today, with so many vehicles having multiple control modules, testing for battery drain will likely involve disconnecting ECMs or isolating circuits.

In older vehicles, disconnect the negative battery cable and connect an ammeter in series with it and the ground post. Make sure the key is off and all loads such as dome lights are turned off. There should be no current draw. If there is, an easy way to isolate the problem is to start pulling fuses one at a time until the draw stops. Another possible draw is through a defective diode in the alternator. Disconnect the positive lead at the alternator to locate this potential problem.

On vehicles with electronic control modules, use a milliamp scale on a DMM and the manufacturer's specifications to ensure that a draw is within parameters (most will draw well under 50 milliamps with the key off). If it measures higher than specified, it might be necessary to disconnect its power supply to be sure that there is not another component in the vehicle causing the additional draw. Be sure to allow time for the ECM to power down and enter "sleep mode" before taking a final reading.

Also, when checking for a battery draw complaint, check surface discharge across the top of the battery (see Task B3).

Task A9   Inspect and test circuit breakers, solid-state current limiters, and fuses; replace as required.

A fuse is an electrical safety device. When it blows, it is because of a current overload somewhere in a circuit. Always repair the problem; never install a fuse of a higher rating. Also, learn to identify the reason for a fuse failure. If the metal filament in the center of the fuse melts, it is caused by a current overload. On older glass style fuses, if the end caps are melted, this is caused by poor or corroded contacts in the fuse holder itself, not a current overload.

A circuit breaker performs the same function as a fuse; however, it has a feature that allows it to be reset after tripping, usually automatically. Most circuit breakers can be identified as a small rectangular box with two studs attached to it. They have their maximum current rating stamped on the housing.

Circuit breaker use is popular because of the possibilities of temporary overloads in bus electrical circuits and the need to restore lost power rapidly, such as for headlights and wipers. A typical circuit breaker is made of a set of contacts controlled by a bimetallic strip of metal. If there is an overload condition, too much current flows through the bimetallic strip and contacts, generating too much heat. The bimetallic strip is made of two different kinds of metal. The uneven expansion of the two metals when heated causes the contacts to open, thus opening the circuit. With no current flow, the bimetal arm cools down, causing the contacts to close and the current to flow through the circuit again. This opening and closing of the contacts will occur over and over until the cause of the overload is determined and remedied. Keep in mind that after repeated cycling, the circuit breaker itself can become weak and oversensitive to heat.

Another common type of circuit breaker for heavy-duty bus applications is the manual resetting type of circuit breakers. The coil type requires the removal of the circuit breaker from power to reset. An overload condition causes the contacts to open and a small amount of current to pass through the coil. The small amount of current is enough to heat up the coil and bimetal arm, keeping the arm open until power is removed. The other style uses a reset button to reset the contacts once the bimetal arm has cooled down from an overload. Circuit breakers are located in a box or panel.

Circuit breakers are handy test devices to have in a tool box. When testing a circuit that continually blows conventional fuses, installing a circuit breaker into the circuit temporarily with jumper wires saves both time and money. Fusible links are short sections of wire designed to melt and open a circuit in case of overload. They are usually installed near a power source (e.g., battery or starter solenoid) and are normally two to four wire gauge sizes smaller in diameter than the circuit they are protecting. When they do melt, the insulation usually bubbles, but not always, making them difficult to troubleshoot. The fuse link has a special high-temperature insulation designed not to separate during an overload. Sometimes the easiest way to test these devices is to simply give them a good tug at either end. If it stretches, it is defective.

To test a fuse or circuit breaker, remove the component and test for resistance using an ohmmeter. A good component should show very low resistance.

A solid-state current limiter (SSCL) is an extremely fast-acting device that provides over-voltage and current spike protection on a power line, especially in cases where high in-rush consumable loads are instantly activated or deactivated. An SSCL also has the capability to protect against voltage drops. Unlike fuses and circuit breakers, SSCLs have the ability to quickly restart after the abnormal condition has cleared to prevent disruptions to surge-sensitive circuits.

Task A10   Inspect and test spike suppression diodes/resistors and capacitors; replace as required.

A diode is simply an electrical "check valve" that allows current to flow in only one direction. The symbol for a diode looks like an arrow with a line drawn perpendicular to its point. A typical use of a diode is in an alternator, where they perform the task of converting AC voltage into DC by simply preventing output to the battery during the negative cycle of the sine wave. Diodes can also be used to prevent back feeding of current from the alternator excitation circuit to key switch circuits once the engine starts.

Testing diodes is simple. You need only to take an ohmmeter and check for continuity through both directions. If it is good, current will flow in one direction but not the other.

Spike suppression devices can be in the form of diodes, resistors, or capacitors. Their purpose is to absorb or redirect a voltage spike that might come from collapsing magnetic field, such as when the AC compressor clutch coil is switched off. By installing such a device in parallel with the coil, the voltage spike is directed back to the clutch coil and prevented from damaging sensitive components, such as ECMs.

Task A11   Inspect, test, and replace relays and solenoids; replace as required.

A relay is defined as a switching device that uses a small amount of current to control a larger one. A solenoid is a device that performs mechanical movement when electrically energized, such as a fuel shutoff solenoid. A solenoid can also incorporate a relay function. A good example of this would be a starter solenoid, which not only moves the starter drive pinion into mesh with the flywheel ring gear, but also makes the high current connection between the battery and the starter field coils.

Most mini-relays have four or five terminals. The two small terminals (often specified as #85 and #86) are used to energize the coil that creates the magnetic attraction necessary to cause a connection between the high amperage switch contacts. Two other terminals make the high amperage connection. These are terminal #30 (the common terminal – "C"), and terminal #87 (the normally open – "NO" – terminal). Sometimes a fifth terminal (marked #87a) is used as a normally-closed contact. Note that the physical size of the power terminals (#30, #87, and #87a) may or may not be larger than the coil terminals (#85 and #86). Amperage capacity will determine terminal size. Also note that some vehicle manufacturers will use a relay in the ground side of a circuit. Instead of terminals #30 and #87 controlling a positive feed, they will control the ground side feed of a circuit.

In operation, when a relay is signaled to close the high-current contacts, a small amount of current is fed through terminals #86 and #85, one being battery positive and the other battery ground. This signal can come from an ECM, key switch, or other low-current switching device. The positive side of the high-current contact (#30) is then connected to the load side (#87), completing the circuit to the high-current device, such as a horn or multiple light circuits. Some relays do not have the same code numbers, but function using the same principles.

Task A12   Read and interpret electrical schematic diagrams and symbols.

You should familiarize yourself with most of the standard symbols for electrical circuit schematics. Valley Forge schematic symbols were once considered standard in the United States and Canada. However, due to increased ownership of trucking and bus OEMs by European companies, some manufacturers are using European schematic symbols and architecture. Schematics are used in ASE tests so you should familiarize yourself with some of the standard symbols. The following figure shows some of the symbols used by truck and bus OEMs, keeping in mind that some manufacturers will use similar, though not necessarily identical, graphics. Pay special attention to the symbols used for grounds, connections, fuses, diodes, switches, relays, and twisted pairs (data bus).

SYMBOL	DESCRIPTION	SYMBOL	DESCRIPTION
⌁	Fuse	▨	Circuit breaker
O▭	Fusible link	▨	Relay
(A1)	Fuse/circuit breaker		
▸▮	Diode	⊥	Switch, push button
⊪⊢	Ground		
◎	Junction point	⌇	Switch, mechanical
J1	Junction point ID	⊘	Switch, Pressure
◉	Light, single filament	⊙	Switch with light
◈	Light, double filament		
⊥	Splice	▨	Sender, oil/water/fuel
⚬⁄⚬	Switch, N.O.		
⚬⚬	Switch, N.C.	⌇	twisted pair
≪	Male/Female connection		
⟨	Female connection		
⟵	Male connection		

Task A13   Read and interpret ladder logic diagrams to diagnose electrical/electronic problems.

Multiplexing is a term used to describe the next generation electrical system. Although there are various manufacturers that offer multiplex systems with different features and approaches, the basic principal of operation is similar. Instead of using individual wires that carry power and ground from various switches and relays to control electrical devices throughout the vehicle, multiplexing simplifies the process by using microprocessors and its own data network to monitor switch positions and distribute power locally to activate electrical devices.

Like other computer-controlled systems, multiplexing uses a program to perform its duties. The program used by typical multiplex systems is called ladder logic. Ladder logic closely resembles a traditional relay-based electrical system schematic with some significant differences. Whereas a hardwired schematic shows the actual flow of battery current, a ladder logic schematic represents instructions written for the software application. Ladder logic can be reviewed or "read" much like traditional hardcopy schematics, or by connecting a laptop or handheld device to the vehicle's multiplex system and viewing the schematics on a monitor.

Read left to right, ladder logic starts with one or more condition instructions or inputs (switch position is on), and ends with at least one control instruction or output (energize light). Each "rung" of the ladder logic program depicts a control circuit (turn interior lights on/off through the interior light switch).

In ladder logic, activation of a device depends on whether the condition of an input is "true" or "false." Continuing with the interior light example, ladder logic is programmed to recognize that when the light switch is closed, a "true" condition exists and power is sent to the interior lights to activate them. The system continually monitors input conditions. When the position of the switch is open, the input condition is "false," so no action is taken and the lights remain off.

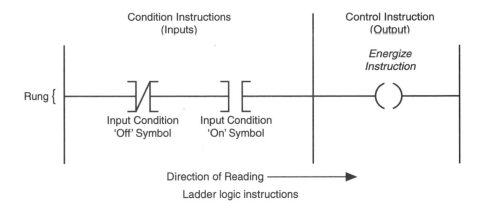

Condition Instructions
(Inputs)

Control Instruction
(Output)

*Energize
Instruction*

Rung {

Input Condition
'Off' Symbol

Input Condition
'On' Symbol

Direction of Reading

Ladder logic instructions

Ladder logic can be programmed to include special control conditions. For example, if an agency wishes to extinguish the first two sets of interior lights to reduce windshield glare for the bus operator at night, ladder logic can be programmed to keep the first two rows of lights illuminated only when the entrance doors are open to allow passengers visibility when entering and existing the bus, and then extinguish the lights after the doors have been closed. In this case, ladder logic will only energize the front two interior lights when both the light switch and door position switch show a "true" condition. If one should show a false condition (light switch is not on or the front door is closed), the two front lights will not be energized.

Task A14   Diagnose and repair computer communication multiplex systems; determine needed repairs.

Essential to multiplexing are various input/output (I/O) modules placed at strategic locations throughout the bus (i.e., above doors, near the engine compartment, etc.) Each I/O module, which contains battery power and ground, is connected to each other by a common data cable and its own data communication network. Despite having its own data communications network, multiplex systems typically can interact with other on-board data networks such as SAE J1708/1587/1939.

When the brake pedal switch at the front of the bus is activated, for example, a traditional power or ground signal is sent to the input side of a module located nearby. However, instead of running hardwires to the back of the bus to activate the brake lights, a data signal is sent to a module located at the rear of the bus, which directs the output side of that module to send power to the brake lights. Because the modules are powered from the bus batteries, multiplex systems are typically programmed to go into a "sleep mode" after a set period of electrical system inactivity to prevent the batteries from draining. Multiplex systems also require a brief amount of time for the data communication to reactivate itself when the system is powered back up.

Multiplexing has reduced the number of wires traveling throughout the vehicle and has greatly simplified diagnostics. The inputs and outputs of the multiplexing modules typically have light emitting diodes (LED) at each location, which provide a simple visual indication if signals are being received or not. When a switch is turned on, for example, the LED at the corresponding input "address" should illuminate.

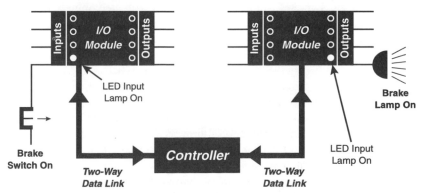

LED lamps provide visual indication of control signals

Connecting a laptop computer or handheld tester to the multiplex system provides another method of tracking faults. A laptop is also used to program the electrical system to add control features (i.e., to automatically activate the headlights when the wipers are turned on), thereby eliminating the need to run additional hardwires and relays as would be done in a traditional electrical system.

Task A15   Using a laptop computer, establish communication with a multiplex control system. Verify that the needed ladder logic inputs are active to control an individual/specific ladder logic output.

In addition to the LED indicator lights that greatly assist in diagnosing multiplexing system problems, there are more advanced equipment and procedures to provide enhanced system monitoring and troubleshooting capabilities. While each multiplexing system offers its own approach, connecting a laptop computer or portable personal computer (PC) contained on a cart with wheels is a common method for establishing a direct "plug-in" connection with the multiplexed electrical system.

Depending on the product, establishing a link with the multiplex system allows many options. Technicians may view ladder logic schematics, check the system's overall functional status, check and verify the function of each multiplex system component including modules, check the input status of each switch or sensor, check output status, and monitor the status of each logic condition. Most systems also have a security feature that only allows authorized personnel to access and operate the system. Other advanced multiplex systems use handheld, wireless devices for diagnostics, troubleshooting, and to control the multiplex electrical system. The multiplex program will not, however, reveal any of the battery voltages present in a given circuit.

Establishing a connection with the multiplex system also allows agencies to program the electrical system to perform control operations that previously were accomplished by adding more wires and relays. For example, if the agency wants the bus headlights to come on automatically when an operator activates the wiper switch, a laptop or other connection to the system allows the ladder logic to be programmed accordingly. While some agencies have the understanding and capability to reprogram ladder logic, others leave this task to the bus manufacturer.

Task A16   Remove, replace, and adjust electrical/electronic switches, sensors, and other electrical/electronic components.

There are many types of electrical and electronic switches and sensors. A sensor is a device that responds to a physical inducement like heat, light, sound, pressure, or motion. Switches are engineered to open and close circuits allowing current to flow when needed to activate a device. The basic is a toggle or on/off switch. This switch operates by physically moving a lever from one position to another (a throw), which will open and close electrical contacts inside the switch. Another basic switch is the pushbutton. This switch works by physically pressing the switch to open and close the electrical contacts inside the switch.

Switches can be engineered to allow current to flow only upon a throw, or when the switch is at rest. Normally open (N.O.) contacts will block current at rest and allow current to flow only on a throw. Normally closed (N.C.) contacts will allow current to flow at rest and block current only on a throw. Switches can also be engineered with a combination of N.O. and N.C. contacts and typically have a common (Com. or C.) contact associated with them. All these contacts will be marked on the case of the switch as N.O., N.C. and Com. or C.

Switches can have more than one set of contacts called poles. Poles isolate one circuit from another while utilizing the same switch. The basic switch would be called a single-pole, single-throw (SPST). Two sets of contacts are called double-pole, single-throw (DPST), three sets three-pole, single-throw (3PST), and so on. Switches having both N.O. and N.C. contacts are called double throws (DT). Double-throw switches in combination with the poles would be called: single-pole, double-throw (SPDT), double-pole, double-throw (DPDT), three-pole, double-throw (3PDT), and so on.

A rotary switch has a knob on a shaft that allows you to select several different positions. A multiple fan speed switch is an example. Another type of switch is the microswitch. The main advantage of a microswitch is that it can be used in tight places where a switch with small packaging is needed. The main disadvantage is that it can only handle small amounts of current. The microswitch is used in places like door interlock controls. Because of the short-range throw of the microswitch, adjustment is critical. A cam switch has a roller lever that rides on a cam. A lobe on the cam causes the switch to throw. A door interlock microswitch is of this type.

A limit switch is a switch that has a lever or a lever with a roller attached at the end of it. These are used in places like a wheelchair lift where the lift platform stowed/deployed limit switch will signal the ECU that it has completely retracted/extended. A plunger-type switch is a momentary switch. The pull-cord passenger stop signal switch, a service interlock switch, and a compartment door light switch are some common examples.

A touch tape and a sensitive mat are other forms of momentary switches. Slight pressure applied connects two copper strips. A touch tape is only about 1-inch wide, but is manufactured in different lengths. It is typically used as passenger stop signals and sensitive edges on wheelchair lifts. Sensitive mats are used to detect the presence of a passenger on the platform of a mobility aid/wheelchair lift or on step treads for automatic passenger counting (APC) systems.

Thermal switches are normally open and close at a preset temperature and are usually stamped on the housing. These are used for engine and transmission hydraulic temperature sensing. Thermal sensors are typically used in HVAC and fire detection systems. Pneumatic or air pressure wave switches can be fixed or adjustable by use of a setscrew. They are engineered by using a contact mounted on a moving diaphragm in response to air pressure. Examples are air pressure switches and passenger door-sensitive edges.

Rheostats and potentiometers are variable resistors, which limit current flow in a circuit. Rheostats have two connections and are typically used in higher current circuits such as instrument panel lamp dimmers. Rheostats are also used as volume control in public address (PA) systems and for gas gauge sending units, which are low current circuits. Potentiometers have three connections and are typically used in low-current signal circuits. A good example is a throttle position sensor. Speed sensors are inductive electromagnetic pickups that produce an AC voltage every time a tooth on a wheel hub moves past the sensor. A feeler gauge is used to set the tight tolerance between the sensor and wheel hub tooth.

A diode is an electronic check valve that permits current to flow in only one direction. The case is marked with an arrow indicating the direction of current flow. These are used as clamping diodes, circuit isolation, or even as a rectifier in an alternator. A transistor is an extremely small electronic switch that contains no moving parts, so it will last a very long time. The disadvantage is that it is very susceptible to high voltage spikes and surges and must be protected. Transistors are used in dashboard indicator diagnostic modules, battery equalizers, voltage regulators, ECUs, and PA systems.

As stated earlier, a relay is a switching device that uses a small amount of current to control a larger one. A traditional relay is an electromagnetic-mechanical device in which the solenoid principle (inductive coil) is applied to the opening and closing of electrical contacts. However, when a relay coil is de-energized, the collapsing magnetic field develops a high voltage spike that can damage or destroy sensitive electronic circuits and equipment. For this reason, a diode (called a clamping diode) is used to suppress the voltage spike. Use of a diode in this type of relay makes it polarity sensitive. As a result, power and ground connections must be properly wired when installing this type of relay or damage can occur. In traditional relays, power and ground could be used interchangeably to energize the relay coil.

A relatively new technology for the transit industry is the collision avoidance system. These systems use different types of sensors to detect objects within a certain distance and provide a warning to the driver. One system utilizes ultrasonic (sound) radar. Another system uses discriminating multiphase microwave radar technology (high frequency radio waves) to detect obstacles.

Configuration of switches

Task A17   ## Use proper care and handling of electrical/electronic components.

Electrical and electronic components found in today's vehicles are designed to withstand harsh environments. The solid-state nature of electronic modules and other components has eliminated moving parts and, for the most part, automotive electronics have become durable and highly reliable. For larger, heavy-duty buses, many agencies specify that electrical system components conform to SAE J1455, "Joint SAE/TMC Recommended Environmental Practices for Electrical System Design." This recommended practice addresses natural and vehicle-induced sources that influence the performance and reliability of on-board electrical and electronic components. Although applicable to Class 6, 7, and 8 diesel power trucks, SAE J1455 has become a part of the Standard Bus Procurement Guidelines developed by the American Public Transportation Authority (APTA).

Today's cars, trucks, and buses are also designed not to generate, or be affected by, electromagnetic interference (EMI) or radio frequency interference (RFI) that can disturb the performance of electrical/electronic equipment as defined in SAE J1113. Most microprocessors have a clock oscillator, and the circuitry uses digital signals for processing and control. These digital signals are square waves, which are ideally suited for digital circuits, but are rich in harmonics that can interfere with radio receivers and other electronics installed in vehicles. In addition, vehicle electronics can also be affected by strong electromagnetic fields. These fields can be caused by nearby transmitters, transmitters installed in the vehicle, and high-voltage power lines. Vehicle manufacturers typically have electromagnetic compatibility (EMC) departments to deal with testing and design issues pertaining to EMI/RFI and to assure compliance with federal regulations and compatibility with factory-installed equipment. It is strongly recommended to consult with the vehicle manufacturer before adding or altering any onboard electronic equipment.

Despite the steps taken by vehicle manufacturers to make electrical and electronic systems rugged, there are certain precautions that must be observed when handling this equipment. Activation arms on microswitches, for example should not be bent in an attempt to avoid adjusting the entire switch for better operation. Likewise, terminal spades or connector pins should not be bent or manipulated in an attempt to obtain a better connection; it is always best to replace the entire connector to ensure continuity. When replacing a wire terminal or repairing wiring, use shrink tubing to eliminate any exposed wiring. Shrink tubing is an insulating "tube" that is slid over the wire or terminal after repaired and then "shrunk" by using a heat gun to provide an electrical insulation.

The introduction of electronic modules on buses and other vehicles prohibits any welding to be done to the vehicle with the batteries connected. Voltage spikes created by the welding process can easily damage electronic equipment. Technicians are urged to read precautions issued by bus and component OEMs before welding on any modern vehicle. A safe approach is to disconnect the batteries along with all electronic components and modules that may be provided with its own internal power source such as engine ECMs, multiplex systems, etc. Also remember to touch yourself to ground to discharge any static electricity when working on electronic equipment.

Shielded data cables are found in multiplex wiring systems and other computer-controlled devices to transmit coded information at high rates of speed. Unlike traditional wiring that carries battery current, these cables are highly sophisticated, and technicians are not advised to make repairs when damaged. Instead, data cables are to be replaced by original equipment manufacturer (OEM) cables when needed.

B. Battery Diagnosis and Repair (6 Questions)

Task B1   ## Perform battery load and/or capacitance tests; determine needed repairs.

To properly perform a battery load test, first determine whether or not the battery is fully charged. There is no point in testing a partially charged battery caused by charging system problems because it will fail. First determine the battery state of charge (see Task B2).

When you are sure the battery is ready to be load tested, first draw off the surface charge if the battery has just been charged (by either an alternator or battery charger). Load the battery by either cranking the starter for 15 seconds or drawing 300 amps with the load tester for 15 seconds. Allow the battery to sit for a few minutes.

To perform the load test, first determine the rating of the battery. This is usually expressed in CCA (cold-cranking amps), although some older batteries might be rated in AH (ampere-hours). Draw the battery down with the load tester at a rate equal to one-half the CCA or three times the AH. Hold this load for 15 seconds. At the end of 15 seconds, with the load still applied, note battery voltage. A reading over 9.6 volts at 70° F means the battery has passed the test. A battery that passes at 10.7 volts versus one that passes at 9.7 volts is the better of the two. One that barely passes will not last long. Keep this in mind in severe weather climates.

Some truck and bus OEMs recommend capacitance testing of batteries and a number of different OEM instruments are available which will perform this test. The test instrument outputs a low potential AC signal and measures the return pulse; the instrument display is idiot-proof and usually reads in terms of OK or not-OK. The danger of using this method is that some good batteries will be rejected as not-OK.

Task B2   Determine battery state of charge by measuring open circuit voltage (OCV) using a digital multimeter (DMM) or perform a specific gravity test using a hydrometer.

A rough estimate of battery state of charge can be determined by measuring its open circuit voltage. With the battery under no load, measure voltage across the terminals. If it is 12.6 volts or more, it is considered to be fully charged. As noted in Task B1, if the battery has been recently charged, either in the vehicle or out, draw off the surface charge using the method described earlier. Before testing, allow the battery to sit for 15 minutes; then, test the open circuit voltage. Any reading less than 12.6 volts indicates that the battery should be charged.

Battery state of charge can also be determined using a hydrometer. A specific gravity reading of 1.265 or higher at 80°F indicates a fully charged battery. A severely discharged battery will show around 1.120 or so. A reading of around 1.200 would indicate a battery that is 50 percent charged. Temperature corrections must be made to the readings for batteries not within 10° of 80°F.

Battery state of charge can also be determined by an open circuit voltage (OCV) test across the battery terminals. When the OCV has stabilized, the following readings can be used:

12.6 volts or more	Fully charged
12.4 volts	75 % charged
12.2 volts	50 % charged
12.0 volts	25 % charged
11.7 volts or less	Fully discharged

Task B3   Inspect, clean, and service battery, cables, terminal connections, and disconnects; replace as required.

As part of any battery service routine, start by wearing eye protection. Inspect the top of the battery case for a buildup of dirt and moisture that can cause a low amperage current draw across the top of the battery. This can be checked by taking one probe of a DMM and dragging it across the top of the battery while holding the other probe on one of the posts. Any significant reading indicates a low current short between the two battery terminals, which can result in a low or dead battery over a period of time. A battery is best cleaned with a water and baking soda solution.

Battery cables and their terminal ends are a frequent source of problems. Many times they are the cause of a no-start or a sluggish starting complaint. A simple voltage drop test (see Task C1) will quickly identify which connection(s) have excessive resistance. Battery tapered-type posts and cable ends are best cleaned with a scraper-type tool that actually peels away all the old corrosion down to bright shiny metal. Flat, screw-type posts are more difficult to clean; however, it is just as important that they be clean in order to transfer current with minimal voltage drop.

When reinstalling battery cable ends, coat the terminals with grease or petroleum jelly, or use a spray marketed for this purpose, to resist corrosion. Protective pads that go under the tapered terminals serve the same function.

When removing a battery and/or cables, always turn off the battery disconnect switch (if provided) and remove the negative cable first, and reconnect it last. This will help to prevent arcing and a possible battery explosion should your wrench come into contact with ground when loosening the positive cable.

If a conventional type battery is low on electrolyte level, add distilled water only. Never top up with acid.

Task B4   Inspect, clean, and repair battery boxes, mounts, and hold-downs; replace as required.

The life of a battery depends in large part on the way it is secured in its mounts. A battery with missing straps or mounts will bounce around and eventually damage the internal separator plates, which may cause an internal short. Similarly, over-tightened hold-down straps may cause the case to crack.

Inspect the battery box when the battery is removed. Clean away any dirt, rust, and corrosion to help combat surface discharge when the battery is reinstalled.

Task B5   Charge battery using slow or fast charge method as appropriate.

The amount of time it takes to properly charge a battery depends on its state of charge. The other factor to consider is whether to charge it at a fast or slow rate. In most cases, a slow charge rate allows for maximum battery life and performance.

When slow charging, a 5–10 amp rate is usually sufficient, although a full charge may take overnight if the battery charge is low. If a fast charge is required, certain precautions must be taken. Never allow more than a 50–60 amp charge rate. Then, monitor the electrolyte temperature to make sure it does not exceed 125°F. If the specific gravity reaches 1.225 during the fast charge, reduce the charge rate accordingly. If gassing or spewing of electrolyte occurs, reduce the charge rate. Never charge a frozen battery until it is brought up to room temperature.

Charge batteries in a well-ventilated area, away from sparks and other sources of heat. If the caps are removed during charging, cover the top of the battery with a moist rag. You can also monitor the specific gravity of the battery during charging to determine its state of charge. A reading of 1.265 indicates a full charge. Alternately, an ammeter on the charger that slowly drops to zero or near zero indicates a fully charged battery. A battery that will not accept a charge from the start (zero amps) is likely to be highly sulfated and should be discarded.

Task B6   Jump-start a transit bus using jumper cables and a booster battery or auxiliary power supply.

Jump-starting a bus or any vehicle with a dead battery can be dangerous. Proper procedures must be followed to ensure that a spark is not generated that can cause an explosion. A battery that is low due to prolonged cranking is likely generating explosive vapors near the vent caps and spark at this location could be dangerous.

Always wear eye protection when jump-starting a vehicle. Then, with both vehicle engines off, connect the positive terminals of both batteries with one cable. Connect the ground booster cable to the negative battery terminal of the booster vehicle, and make the last connection to the frame or chassis (ground) of the dead vehicle. This will prevent sparks in the area of the dead battery.

Start the engine of the booster vehicle. If the jumper cables are of a generous size, crank and start the dead engine immediately. If the cables are small, allow the connection to remain for a few minutes while the dead battery recharges. Then, with the help of the booster battery, crank and start the engine. Remove the cables in the reverse order of installation.

C. Starting System Diagnosis and Repair (8 Questions)

Task C1   Perform starter circuit voltage drop tests; determine needed repairs.

Cranking system-related problems are common vehicle malfunctions. It is important to know how to perform a few simple tests in this area to quickly and effectively solve the problem. A cranking system voltage drop test identifies high resistance in the cranking circuit that can cause a slow or no-start condition. Every connection and conductor in the circuit, from the battery to the solenoid through the ground path in a potential problem area.

To test for voltage drop in the positive battery cable, set a DMM to the V-DC scale and attach one probe to the positive terminal post of the battery (not the cable or terminal). Touch the other probe of the DMM to the starter solenoid stud (again, not the cable itself). While holding the probes in this position, crank the engine until you get a steady reading. Ideally, there should be less than a half-volt drop. If there is more, narrow the problem down by checking point to point along the cable (e.g., battery post to cable end, cable end to cable wire, wire end to wire end). Follow the same procedure noted earlier. (Polarity with a DMM is not important. If the leads are backward, the display will show a minus sign in front of the reading.) Ideally, each connection along the cable should have less than one-tenth of a volt drop.

Another way to check for high resistance is to crank the engine for several seconds and then feel along all the connection points and conductors for heat. If any point is more than just warm, there could be excessive voltage drop at that location. Clean or replace as required.

Task C2   Inspect and test components of the starter control circuit (master/key switch, pushbutton and/or magnetic switch, and wires); replace as required.

The starter control circuit includes those components between the starter switch (Note: the starting switch can be key or pushbutton activated depending on the bus type) and the magnetic switch (or starter solenoid itself if there is no magnetic switch in the circuit). It is important to remember that this is a critical, high-current circuit. Current requirements of solenoids on some larger starters are upwards of 100 amps, often necessitating a separate relay (also known as a magnetic switch) to handle the high amperage load. Magnetic switch assemblies are typically 4-post units, having two large and two small terminals. However, some can also resemble large 5-pin mini-relays.

The starting switch is a low-current switching device that controls the magnetic switch (relay). Current from this relay then goes to the starter solenoid (another relay) to engage the starter. Both the control and light current circuits need to be in working order to get the engine to crank. Also, between the starting switch and the magnetic relay are sometimes found neutral safety switches, designed to prevent an engine from starting in gear.

Some manufacturers do not use a separate magnetic switch between the starting switch and the starter solenoid, which makes the key switch circuit moderately high-current. Add a neutral safety switch and some bulkhead harness connections, and there is plenty of potential for trouble in the form of a no-start condition if everything is not in good working order.

It is best to have the wiring diagram for the vehicle to effectively troubleshoot the starting control circuit. Properly understanding relays (see Task A11) and voltage drop testing (see Task C1) are essential for diagnosing starting circuit problems.

Task C3   Inspect, test, and replace starter relays and solenoids/switches; replace as required.

Most starter solenoids have two functions: they shift the starter drive pinion into mesh with the flywheel ring gear, and make the high-current connection to deliver battery power to the starter motor field coils.

If a starter fails to crank, and all the other circuits to the solenoid check out, it is important to remember the set of electrical contacts in the solenoid that are subject to pitting and corrosion over time. Testing them is easy. Perform a voltage drop test across the two large posts on the solenoid while cranking the engine, just as you would for a battery cable or clamp.

If a starter pinion fails to engage the ring gear, the cause could be a solenoid fault. When replacing a solenoid, keep in mind that quite often a pinion clearance adjustment needs to be made at the same time.

Task C4   Remove and replace starter; inspect flywheel ring gear or flex plate.

Whenever removing a starter motor, inspect the teeth on both the starter drive pinion and the flywheel ring gear. Compare with new components if unsure of what is worn. Keep in mind that both the pinion and ring gear have a machined chamfer on the teeth to facilitate engagement.

If you determine that the flywheel ring gear needs to be replaced, see the engine manual for the appropriate procedure. If the starter pinion is worn, it can sometimes be replaced economically compared to the cost of replacing the whole starter. However, take into consideration the age and condition of the starter motor itself before attempting a pinion drive replacement.

Task C5   **Inspect, clean, repair, or replace cranking circuit battery cables and connectors.**

Testing of starter circuit components using the voltage drop method was discussed in Task C1. This is the preferred method for determining the serviceability of cranking circuit components. Obvious physical damage such as cables with rubbed insulation and clamps with broken or cracked ends warrants replacement.

If the battery cable is OK, but needs new terminals, avoid using generic bolt-on cable terminals. They do not have the current-carrying capability that crimp- or solder-type terminals have, and will almost certainly cause trouble later with corrosion and high resistance. Use the appropriate size crimp or soldered terminal along with shrink tubing when repairing battery cables to minimize corrosion and resistance.

Task C6   **Differentiate between electrical, multiplex, or mechanical problems that cause slow cranking, no cranking, extended cranking, or a cranking noise condition.**

There are many causes of slow crank, no crank, extended cranking, or cranking noise conditions, and determining whether the cause is related to the engine or the electrical cranking circuit can be challenging to the technician. When such problems occur, the technician must evaluate whatever evidence is apparent by listening, checking that electrical connections are not overheating, and observing smoke emission during cranking that would indicate that fuel was being injected. The guided diagnostic software used by some engine manufacturers is designed in a systematic manner to isolate electrical, electronic, and mechanical problems. Use it when it is available. In most cases, this software is designed to rapidly source electrical and fuel system problems, and can additionally direct the technician to locate engine mechanical problems. Manually turning an engine over will quickly verify whether an engine is mechanically seized or not. Failing to do so may result in extensive damage, including the possibility of a fire because the starter motor can overheat as it tries to turn over a seized or "frozen" engine.

A possible starter problem is one where the starter engages and immediately disengages. The problem is created when an open develops in the starter solenoid's hold-in coil (also referred to hold-in winding). When this occurs, the pull-in coil (also referred to pull-in winding) engages normally, and then loses its ground and magnetic field when power flows through the starter motor's field coil. With an open in the hold-in coil, however, the magnetic field needed to keep the plunger engaged cannot be maintained. Since the pull-in coil is now at rest, the plunger will immediately disengage. The process keeps repeating itself as long as power is present at the switch terminal (SW) on the starter solenoid.

D.   Charging System Diagnosis and Repair (9 Questions)

Task D1   **Diagnose the cause of a no-charge, low-charge, or overcharge condition; determine needed repairs.**

A no-charge complaint could be caused by any number of different reasons, from a simple blown fuse to faulty brushes or diodes inside the alternator. Due to the many different types of systems, it is impossible to list all the potential causes. Consult the service manual for the system you are working on and follow the troubleshooting procedures.

A low-charge condition is one where the alternator charges properly with light loads, but falters under heavy ones.

An overcharge condition (system operating over the maximum specified voltage) usually indicates a defective voltage regulator, but not always. In some systems, a defective sensing diode in the alternator can send a low signal to the regulator, forcing it to overcompensate.

Task D2 **Inspect and adjust alternator drive belts/gears, pulleys, fans, mounting brackets, and tensioners; replace as required.**

While worn drive belts are visually obvious, try to determine the cause of the failure. Two of the more common causes are incorrect belt tension and misalignment. On vehicles without automatic tensioners, recheck belt tension after installing a new belt. Allowing the system to run for a while will seat the new belt, after which readjustment is required. A loose belt will slip and fail prematurely.

Some alternators are shimmed fore and aft in their mounting brackets. Be sure to align the alternator-driven pulley with the engine drive pulley.

Task D3 **Perform charging system output tests (12 volt and 24 volt); determine needed repairs.**

Alternator output tests are simple procedures. All that is required is a battery load tester and a voltmeter. Before starting, look up the model ID on the alternator and check the specifications for maximum output. Also, check the drive belt for proper tension.

Attach the battery load tester across the battery terminals as if you were going to load test it. However, for this test, attach the amp clamp (either the one with the load tester or a handheld meter) around the alternator output wire. With the engine at high idle, load the battery tester down to a value slightly higher than the alternator rated output. This forces the alternator to output maximum amperage, which should be within 5 percent of specs. Note, however, that with the load tester drawing more current than the alternator can replace, system voltage will be down. This is normal. Next, reduce the current draw on the load tester to an amount slightly less than the alternator rated output. System voltage should then increase to over 13.5 volts.

Some systems can be tested for maximum output by a procedure known as full-fielding. NOTE: Full field testing is not recommended because unregulated voltage can damage electrical and electronic components. If you are full-fielding an alternator, follow the manufacturer's instructions carefully.

Task D4 **Perform charging circuit voltage drop tests; determine needed repairs.**

For the alternator to provide maximum output, the output wire must be in good condition as well as the ground of the alternator. Check both by testing voltage drop in each circuit.

Voltage drop in the alternator output wire can be tested in one of two ways. The first method involves performing the maximum output test (Task D3), and then measuring voltage drop between the output stud on the alternator and the positive battery terminal. Perform this test using the same method as the starter circuit tests noted in Task C1. The reading should be less than one-tenth of a volt. Anything more indicates high resistance between the two points.

With the alternator at maximum output, test for voltage drop between the alternator housing and the battery ground terminal. Again, this should be one-tenth of a volt or less.

A second method of testing voltage drop in the alternator output circuit can be done with the engine not running. Connect the positive clamp of the load tester to the output stud on the alternator. Connect the negative clamp to the case. Draw current through the load tester equal to the specified alternator maximum output. Do not exceed this amperage, or you may melt a wire. Measure voltage drop as described earlier.

Task D5 **Test, adjust, or replace voltage regulator.**

The alternator, which is typically belt-driven from the engine, produces the electricity needed to charge the battery and to operate electrical equipment. The output of these devices, however, continues to rise as engine speed increases and can at times exceed the battery's charging requirements. The role of the voltage regulator is to control the field current applied to the alternator. When there is no current applied to the field, there is no voltage produced from the alternator. In a 12-volt system, the voltage regulator typically maintains an output of about 14.5 volts to keep the battery fully charged. When the voltage exceeds 14.5, the regulator will stop supplying voltage to the field and the alternator will stop charging. In some applications, the voltage regulator is integral with the alternator. In more traditional applications where the voltage regulator is a separate device, it is important that the integrity of all wire and terminal connections between the two devices be maintained.

Early voltage regulators were electromagnetic devices with wire-wound coils, contact points, and bimetallic hinges. They were replaced with more reliable solid-state regulators, which were also less affected by temperature changes. In more advanced applications, the voltage regulator has become part of the engine's electronic control system.

Task D6   Maintain, remove, and replace alternator.

Replacing an alternator is a relatively simple job. Label wires before removing them. Be sure to disconnect the battery cut-off switch (if provided) and the negative battery cable before removing the alternator wiring and after reinstalling the new alternator to prevent accidental sparks and possible wiring damage. Properly torque the fasteners and ensure the pulleys are aligned.

Task D7   Inspect, repair, or replace charging circuit connectors and wires.

In order for the alternator to transfer all of its power efficiently to the battery and the various loads in the vehicle, the wiring attached to the alternator must have clean connections and be in good condition. Other than a visual inspection for chafes, corrosion, etc., the best way to test the output wire is with the voltage drop test discussed in Task C1.

When repairing wiring in the charging system, use the proper size cabling and connectors for the circuit you are working on. Charts are available which list the recommended wire gauge size depending on amperage flow, length of wire, and voltage. Also, use the proper crimp tools along with shrink tubing to ensure a good connection.

Task D8   Check battery equalizer output, check wiring and mounting; determine needed repairs.

A battery equalizer is typically used in larger buses with a 24-volt battery system where some of the electrical components such as lights and accessories such as the farebox need to operate on 12 volts. The equalizer allows 12 volts to be drawn from one 12-volt battery while dividing the current load across all batteries in the bank. The equalizer is designed to keep the entire battery bank balanced to prevent overcharging and boiling of batteries. Equalizers continuously monitor battery condition whether the engine is running or not. Most models provide short circuit protection and include a re-settable circuit breaker to protect against overloads and reversed polarity conditions.

A battery equalizer is connected at the +24 volt, +12 volt, and ground points of the battery system. When a 12-volt load is present at Battery A and the voltage of that battery drops to a level just below that of Battery B, the equalizer activates and transfers current from Battery B to Battery A to satisfy the load. The equalizer will even-off the charge when the load is reduced. Equalizers are available with various amp ratings.

Most equalizers can be mounted in any orientation. However, because they generate heat, the recommended orientation is vertical with adequate clearances and ventilation provided in the area housing the equalizer to allow for heat dissipation. Equalizers also need to be protected from rain and moisture, and must be mounted in such a way as to prevent metal objects from shorting the terminals.

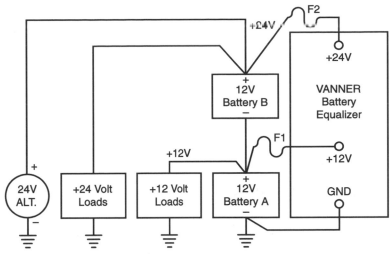

Note - Battery Banks A and B should
have the same amp-hour capacity.

Task D9   Verify operation of charging system circuit monitor; determine needed repairs.

The charging system circuit monitor consists of a diagnostic module and one or more relays that connect to the R (relay) terminal on the 24-volt generators found in larger buses. The output on the R terminal is one-half of the generator output. This output is then fed to a 12-volt coil relay, which drops out below approximately six (6) volts. In cases where a 12-volt alternator is used, a failure is detected when the voltage drops below 12 volts. Twelve-volt alternators normally operate with a charging output between 13.6 and 14.4 volts. This reduction or loss of output voltage causes any other relay in this system and the diagnostic module to indicate a generator or alternator problem. This circuit can be wired to shut down nonessential systems like the HVAC when the generator fails.

The diagnostic module could be a diode module, a transistor module, or part of the multiplexing I/O circuit. These modules allow the testing of the instrument panel indicators and alarms by utilizing a test switch on the dashboard. The diagnostic module isolates the indicators and alarms from each other and their related circuits. A diode module contains all diodes. A transistor module contains diodes and transistors and may contain an indicator dimming circuit for "nighttime" running. As part of the multiplexing I/O circuit, the indicators can be programmed for dimming, flashing, and even used as an audible warning signal to alert the bus operator.

E. Lighting Systems Diagnosis and Repair (8 Questions)

Task E1   Headlights, Daytime Running Lights, Parking, Clearance, Tail, Interior, and Dash Lights (5 Questions)

Task E1.1   Diagnose the cause of brighter than normal, intermittent, dim, or no headlight and daytime running light (DRL) operation.

Multiple lights that are brighter than normal can only be caused by an alternator that is overcharging. Verify this by checking alternator output with a DMM.

A malfunctioning charging system can also cause dim light operation if chassis voltage is too low. Verify this problem with a DMM also.

Dim lights can also be caused by problems such as excessive resistance in fuse holders, relays, wiring, switches, connectors, and chassis grounds. Of all the aforementioned possibilities, suspect the chassis grounds first. Poor grounds cause a majority of problems related to this complaint, from simple loose hardware to poor metal-to-metal contact between chassis components. Voltage drop tests (see Task C1) between the ground side of the bulb and the battery negative post will confirm ground integrity problems.

Intermittent operation can be caused by a cycling circuit breaker or a loose connection somewhere in the circuit. Often circuit breakers are incorporated directly into the headlight switch. A cycling circuit breaker is caused by either a defective breaker or an overload in the system, and will result in on-off circuit operation. Loose connections are harder to find. A good tip would be to gently pull and wiggle suspect harnesses and connectors while watching the light action.

A blown fuse, defective circuit breaker, bad switch, or an open in the wiring generally causes no light operation. At this point, it is best to get a wiring diagram for the vehicle and probe with a digital multimeter (DMM) at various points downstream from the power source until you find the open.

Daylight running lights (DRL) are now mandated in some jurisdictions and in all of Canada. Some OEMS specify DRLs as a default option and increasingly, fleets are using them. DRLs are a safety feature and illuminate anytime the vehicle is running without the vehicle headlights being switched on.

Task E1.2   Test, aim, and replace headlights.

Headlight aim should be checked on a level floor with the vehicle unloaded. In some states, this may conflict with existing laws and regulations. If so, modify the instructions to meet the state's legal requirements.

To adjust headlights, first check headlight aim. Various types of headlight aiming equipment are available commercially. When using aiming equipment, follow the instructions provided by the equipment manufacturer.

When headlight aiming equipment is not available, aiming can be checked by projecting the upper beam of each light upon a screen or a chart at a distance of 25 feet ahead of the headlights.

Some manufacturers recommend coating the prongs and base of a new sealed beam with dielectric grease for corrosion protection. Use an electrical lubricant approved by the manufacturer.

Sealed-beam halogen and xenon headlights are designed to give substantially more light on high beam than incandescent, extending the driver's range of visibility for safer night driving. Both halogen and xenon bulbs produce a whiter light, which helps improve visibility. They also last longer, stay brighter, and use less wattage for the same amount of light produced. When replacing individual replaceable bulbs, avoid touching the glass envelope. Oil from the skin can cause the bulb to shatter when turned on.

To obtain maximum illumination, headlights are required to be kept in adjustment. Not only is proper aiming important for good light projection onto the road, but also discomfort and dangerous conditions can be created for oncoming drivers if the headlights are not properly aimed. Two methods that can be used to check headlight aiming are: the screen method and the portable mechanical aiming unit method. Regardless of the method, the vehicle's tire inflation, springs, and proper ride height should be checked first.

Task E1.3   Test headlight and dimmer switches, wires, connectors, terminals, sockets, relays, and control components; repair or replace as required.

Headlight switches are used to control the operation of the headlights and sometimes the parking and dash lights as well. Some manufacturers use separate toggle or rocker switches for each function, while others incorporate all switching functions in one multifunction switch. Some bus manufacturers consolidate the headlight switch into the Master Run Switch, which also controls the ignition circuit. It is important to understand how a particular system is constructed to make troubleshooting easier. Always consult the wiring schematic for the vehicle you are working on.

Dimmer switches switch current flow between the high and low beam circuits. This switch can either be floor mounted or incorporated into a multifunction turn signal switch sometimes known as a stalk switch. The dimmer switch is a simple device that directs voltage to either the high or low beams. One wire is for power in; the other two direct current to either the high or low beam circuits.

Many headlight circuits contain relays to reduce the load conducted through the switch itself. A defective headlight or dimmer switch should affect both left and right headlights because they are wired in parallel. However, if only the high or low beam circuits are not operational, the headlight switch can usually be ruled out, because it is the dimmer switch that distributes power to the lights.

A dimmer switch is connected in series with the headlights to control the current path for high and low beams. In buses, it is either located on the floorboard or on the steering column. Testing of the switch is fairly straightforward. One method is to look for voltage at the battery or power terminal and then look for voltage at the high- or low-beam terminal when the switch is switched. Another method is to use a jumper wire to bypass the switch. If the headlights operate with the switch bypassed, the switch is faulty. Older vehicles tend to have problems with the floor-mounted switches due to their location, which subjects the switch and connector to rust, corrosion, etc.

Task E1.4   Inspect, test, and repair parking, clearance, and taillight circuit switches, bulbs, sockets, connectors, terminals, relays, wires, and light-emitting diodes (LEDs); replace as required.

Parking, clearance, and taillight circuits can be controlled by a Master Run Switch, a multifunction headlight switch, or they can be powered by separate toggle or rocker switches. When troubleshooting these lights, keep in mind that the most common causes of trouble are usually related to poor grounds, either at the lights themselves, or at a ground strap.

If only one clearance, taillight, or parking light is dim and the rest are OK, suspect a problem with that particular light or its ground. If all of the lights are dim, then assume that a ground or power supply malfunction is the cause.

Light-emitting diode (LED) lighting units are being increasingly used on buses. Despite higher initial costs, LEDs produce no heat and last much longer with less maintenance. Use OEM instructions to test but bear in mind that these are diodes so polarity is important when testing them.

Task E1.5   Inspect, test, and repair dash light circuit switches, bulbs, sockets, connectors, terminals, wires, and printed circuits; replace as required.

Dash illumination lights (not to be confused with the warning lights) on most buses are fed from the same power source as the taillights and clearance lights on the vehicle, usually through the multifunction light switch assembly. However, between this source and the lights themselves will usually be a rheostat, also known as a dimmer switch, to allow the driver to reduce the intensity of the lights when driving at night. This dimmer switch may either be incorporated into the headlight switch, or be remote mounted elsewhere on the dash.

If the dash lights on a bus are dimmer than normal and all the other lights work OK, suspect a problem with the dimmer switch or its wiring, assuming that it has been adjusted to the correct position.

Task E1.6   Inspect, test, and repair interior and exterior light circuit switches, bulbs, sockets, connectors, terminals, ballasts/inverters, and wires; replace as required.

The most common interior lights found in smaller, automotive-style buses are the dome and courtesy lights. The typical dome courtesy light can be operated through a switch that is driver controlled. Most dome lights also operate automatically whenever a door is opened through the use of a door jamb switch. These are internally grounded switches located in one or both of the doors. Opening the door causes the switch to close and the dome and courtesy light(s) to illuminate. Some dome lights have a switch incorporated into the dome light assembly. That way if the doors are closed, the dome light can still be activated through the use of the dome light switch. Luggage compartment lights can also be activated using a door jamb-style of switch or through the use of mercury switches located on the luggage doors. Opening the door causes the switch to close and the lights to come on. Panel lights can be controlled through the use of the headlight switch, separate toggle switch, or a solid-state panel light switch. Most panel light switches or headlight switches permit the driver to adjust the brightness of the panel lights. A rheostat is used to vary the resistance of the current flow, thus controlling the brightness of the panel lights.

Larger, full-size buses also use fluorescent lights to provide interior passenger lighting. With a fluorescent light, electricity is delivered to a ballast, which sends a spark through the mercury-vapor-filled tube (bulb), creating light by activating the phosphors that coat the inside of the tube. If any of the components or wiring are faulty, the light will not function. Other problems could be caused by poor contact between the pins at the ends of a lamp tube and the lamp sockets.

Some fluorescent lamps are quieter than others, but most have a slight hum. If the sound seems too loud or there is an electrical burning smell, the ballast is probably either the wrong type, improperly installed, or defective. The best way to diagnose defective ballasts is to make sure that all other components and wiring connections are intact. Check for power and ground at the lamp ballast and lamp sockets, check that both lamp sockets are intact, and check that the lamp is not burnt out. If all is in order and the problem still exists, then replace the ballast.

Task E2   Stoplights, Turn Signals, Hazard Lights, and Backup Lights (3 Questions)

Task E2.1   Inspect and test stoplight circuit switches, bulbs, sockets, connectors, terminals, relays, control components, and wires; repair or replace as required.

Stoplight circuits are relatively simple in that the major component is the switch itself. These are two types. On buses with hydraulic brakes, the switch is usually located on the brake pedal, activated by simple mechanical movement. On buses with air brakes, the switch is incorporated into one of the brake application air lines. Air pressure acting against a diaphragm will close electrical contacts in the switch to complete the circuit. Some buses will have three such switches: two for service brake applications (one for primary, one for secondary), and another for parking brake applications.

In some bus applications, stoplight and turn signal lamps share the same bulbs, current from the stoplight switch is usually routed through the turn signal switch. This allows it to be directed properly when both the brakes and turn signals are activated at the same time.

Task E2.2   Diagnose the cause of turn signal and hazard flasher light system malfunctions; determine needed repairs.

The turn signal circuit directs output from the flasher unit to one side of the bus or the other depending on the position of the switch. If one side works properly but the other does not, assume that the flasher unit is OK and that the problem lies in the malfunctioning circuit. The brake lamp switch can also be considered part of the turn signal circuit, because if the brakes are applied during a turn, the direction the bus is being turned to must continue to flash. This is accomplished by routing the current from the stoplight switch directly into the turn signal switch. At this point, the turn signal switch will direct the current flow in the proper direction depending on position.

Task E2.3   Inspect and test turn signal and hazard circuit flashers or other control components, switches, bulbs, sockets, connectors, terminals, relays, wires, and light-emitting diodes (LEDs); repair or replace as required.

Turn signal flashers that operate faster or slower than normal can indicate higher or lower-than-normal current load on the circuit. For example, if a bulb burns out on one side of a vehicle, this would reduce current demand in that particular circuit, usually causing the flasher assembly to blink slower than normal. Conversely, if a condition caused higher-than-normal current flow in the circuit, due to extra lights being added, current flow would be increased, causing the bi-metal contacts inside the flasher to heat up and cycle at a faster rate. This condition could cause premature failure of the flasher assembly.

Newer vehicles with electronic flashers will incorporate a relay in the circuit to handle the high-current switching demands. This type of system is not as affected by either defective or additional lights in the circuit.

Task E2.4   Inspect, test, and adjust backup light and warning devices, circuit switches, bulbs, sockets, connectors, terminals, and wires; repair or replace as required.

Switches mounted on the transmission itself or a signal sent by the transmission's ECU usually control the backup circuit. When the transmission is shifted into reverse, contacts inside the switch are closed or a signal from the ECU is sent to the relay circuit and the backup lights are energized.

Warning alarms are installed on most buses to alert persons near the vehicle that it is about to back up. These devices use the same switch and signal that the backup lights do, but also incorporate a relay in the circuit to handle the high power demands of the alarm.

F. Gauges and Warning Devices Diagnosis and Repair (4 Questions)

Task F1   **Diagnose the cause of intermittent, high, low, or no gauge readings; determine needed repairs.**

When troubleshooting gauge systems, first determine whether the problem exists with all the instrumentation or just isolated gauges. For example, older style bi-metal gauges are sensitive to voltage fluctuations. If you find that they all read too high or too low, it is possible that the instrument voltage regulator is malfunctioning. The purpose of the instrument voltage regulator is to maintain a steady voltage supply to the gauges, regardless of battery voltage. Individual gauges that read erratically may be caused by any number of factors, including high resistance, a defective sending unit, or a malfunctioning gauge. Follow manufacturer's instructions for out-of-dash gauge testing.

Later model instrumentation systems primarily use magnetic-style gauges. Most are not affected by changes in system voltage, and therefore do not use a separate instrument voltage regulator. Follow OEM instructions when diagnosing electronic dash displays.

Not all gauges in an instrument panel are electrically operated. Some gauges, such as air application gauges, are mechanically actuated and not affected by electrical problems. Air application gauges must test within 4 psi of an accurate master gauge.

Never rely only on the accuracy of dash gauges to make major decisions such as determining the need for an engine rebuild. For example, if vehicle instrumentation indicates low engine oil pressure, confirm the problem with a known accurate, mechanical gauge to verify the condition.

Task F2   **Diagnose the cause of control area network (CAN) driven gauge malfunctions; determine needed repairs.**

Instrumentation systems in current vehicles rely on information sourced from sensors that transmit information to many vehicle control system modules including the engine, transmission, and antilock brake system (ABS). Each module can broadcast information over a control area network, referred to here simply as a data network or "network," to the other on-board computers. While older heavy-duty vehicles use the SAE J1708/1587 drivetrain data network, newer vehicles use the faster SAE J1939 network. One computer is usually dedicated to instrumentation. It is the function of the instrumentation module to receive digital signals from the chassis data network and display them on the dash in analog, bar graph, or digital format, depending on vehicle options.

Newer instrumentation systems will go into a self-test mode when the master switch/key is first turned on to verify the operation of each gauge. When the accuracy of any gauge is suspect, first determine if the dash is receiving accurate information. To do this, connect the appropriate electronic diagnostic service tool or "reader" tool to the vehicle diagnostic connector and scroll the menu options to scan the various sensor values that the module is receiving. From there you can determine the problem source by checking the values against actual vehicle operating parameters.

Task F3   **Inspect, test, and adjust gauge circuit sending units, sensors, gauges, connectors, terminals, and wires; repair or replace as required.**

To test sending units, use the diagnostic tool or manufacturer's service manual for the vehicle to determine what the senders should read at specific operating parameters. For example, a certain resistance value might be specified for a fuel level sending unit when the tank is half full. Or, a temperature sending unit could be tested in boiling water, again, with a specified resistance value at that point. Another quick way to check thermistor-type temperature senders on an electronic vehicle is to compare values (using a diagnostic tool) to those specified in the OEM software parameters.

Many electrical gauges and some electronic ones can be checked using a variable resistance test box. Unplug the wire at the sending unit and substitute the resistance of the test box while comparing the action of the gauge against manufacturer specifications.

Task F4   **Inspect and test warning device (lights and audible) circuit sending units, sensors, bulbs, audible components, sockets, connectors, terminals, wires, and printed circuits/control modules; repair or replace as required.**

Warning lights and buzzers on older vehicles (pre-electronic) have a sender or switch on the ground side of each monitored function that will close and allow the circuit to be powered up in the event of a malfunction.

Newer vehicles have built-in warning systems designed to alert the driver in case of a malfunction that might damage the engine or other components. Depending on the fault, the engine may reduce power or shutdown. Usually, an audible alarm unit is located behind the dash and will sound when a malfunction is detected. Often they are incorporated into the dash itself and not serviced separately; however, some are remote mounted and can be replaced.

Buses with electronically-controlled engines typically have at least two dash warning lights to indicate faults. One will be a "check engine" light, and the other a "stop engine" light. The former will alert the driver to a condition that needs attention at the driver's earliest convenience, while the latter will indicate a problem that requires immediate engine shutdown, such as low oil pressure or high water temperature. Depending on how the engine protection system has been programmed, the stop engine light alert may be accompanied by either a ramp-down in engine power and/or an automatic engine shutdown.

When testing circuits in an electronic instrumentation system, always use a DMM to check voltage and never test lights. Test lights draw too much current from an electronic circuit to make this type of testing valid and may damage the circuit. It should be noted that most electronic circuits operate at voltages lower than battery voltage.

Task F5   **Inspect and test electronic speedometer and odometer systems; replace as required; verify proper calibration for vehicle application.**

Most speedometers and odometers on current buses are driven electronically by means of information received off the data network (see Task F2). When the information on any of these indicators is suspect, first check the calibration menu of the manufacturer diagnostic software tool to see how the vehicle speed parameters are set up. For example, in order for the computer to accurately display speed or mileage data, it must first know the rear axle ratio and tire size. Some engine/vehicle management systems even track tire wear to adjust the speedometer reading. If these parameters are not properly programmed into the computer, inaccurate data will result. These data fields may be password protected.

Many speedometers get their information through sensors known as magnetic pickups. These are simple devices that contain a permanent magnet wrapped by a coil of wire. They are solidly anchored to the axle at the wheel end and are located close to a rotary-toothed wheel. As the wheel rotates and the teeth cut the magnetic field, a small AC voltage signal is generated. Signal frequency is used by the ECM to determine engine or wheel speed.

G. Related Systems (5 Questions)

Task G1   **Inspect and test horns, horn circuit relays, switches, connectors, terminals, and wires; repair or replace as required.**

When testing a horn relay circuit, first check for fused power at the positive side of the horn relay (usually pin #30). Next, check for a signal from the steering wheel switch. This may be either positive or ground side switching, depending on the manufacturer. Assuming the use of a 5-pin mini-relay, the circuit should function exactly as described in Task A9.

Task G2   **Diagnose the cause of constant, intermittent, or no wiper operation; diagnose the cause of wiper speed control and/or park problems.**

Larger buses sometimes use air-driven wipers with two motors or electric-driven wipers with one motor, while smaller buses typically use electric-driven wipers. Most electric windshield wiper systems use multispeed motors. The motor will either be a permanent magnet type with high and low speed brushes, or it will have an external resistor pack much like heater blower motors. Inside the first style of motor will be a low speed and a high speed brush set. If the motor works sufficiently in one speed but is sluggish or not functioning in the other, it may be due to defective brushes inside the motor. First, ensure that the proper voltage is coming from the appropriate side of the wiper switch with the circuit closed. Consult the appropriate manufacturer's repair manual for the wiring diagram of the vehicle.

If a motor fails to work at all with power applied, note that some motors incorporate thermal overload protection should a motor overheat due to binding linkage.

Task G3   **Inspect and test wiper motor, resistors, park switch, relays, switches, connectors, terminals, and wires; repair or replace as required.**

If wipers should fail to park properly after the switch is turned off, it may be due to a defective park switch usually incorporated into the motor itself. The function of this switch is to continue motor operation until the blades are fully retracted. Also, check that the park switch is receiving power when the wiper switch is in the OFF position.

Task G4   **Inspect and test windshield washer motor or pump/relay assembly, switches, connectors, terminals, and wires; repair or replace as required.**

Most washer pump circuits are relatively simple ones. The switch for the pump is usually incorporated into the wiper control switch, and will control the washer pump through either positive or ground-side switching. If the wipers work but the washers do not, you can eliminate a fuse or circuit breaker problem since the wiper switch assembly usually draws power from the same source.

Task G5   **Inspect and test side view mirror motors, heater circuit grids, relays, switches, connectors, terminals, and wires; repair or replace as required.**

Movable electrically-powered mirror assemblies incorporate two different motors—one for horizontal movement and the other for vertical. If one or both motors do not operate, first confirm that power exists at the motor terminals when the appropriate side of the mirror switch is depressed.

Heated mirrors can be considered a safety option in cold climates. When testing a heater circuit for proper operation, be aware that some circuits contain a fast warm-up cycle, which then drops considerably in current draw as the mirror warms.

Task G6   **Inspect and test HVAC electrical components including: A/C clutches, motors, resistors, relays, switches, controls, connectors, terminals, and wires; repair or replace as required.**

Heater blower motors typically have multiple speeds. To accomplish this, the current flow from the switch is routed through a set of resistors wired in series. If one of these resistors should fail, the motor may work in high speed only, since the resistors are bypassed in the high-speed position. Blower motor resistors are usually found in the blower air stream to keep them from overheating. Even though this voltage drop is intentional, it still produces heat.

A/C compressor clutches can be controlled by either a set of manual switches and pressure sensors, or directly from an A/C system microprocessor. Research how the system is configured before troubleshooting.

When replacing A/C compressor clutches, many need to be shimmed properly to maintain proper air clearance in the disengaged position. This must be done carefully. Too little clearance may cause the clutch to drag and burn up; too much might prevent engagement.

Task G7   **Inspect and test engine cooling fan electrical control components; replace as required.**

Some buses are equipped with electrically-driven engine cooling fans, while others have a hydraulically-driven and electronically-controlled cooling fan. This type of system typically consists of a fluid reservoir, pump, spool/switching valve, electronic control valve or electric solenoids, fan motor, and piping. The operation of the fan motor is basically just the opposite of that of the hydraulic pump. Fluid under pressure is released through the two pump gears of the motor. The pressure is dissipated while causing the gears to turn the output shaft. The fan blade, which is attached to the output shaft, rotates and cools the coolant flowing through the radiator core.

Fan speed is controlled by the electronic control valve based on signals received from the engine's electronic control module (ECM). The ECM monitors coolant, charge air, and engine oil temperature, and activates the fan when pre-defined temperatures are exceeded. As an engine protection feature, the fan will automatically go into high-speed mode if battery current is not present with the engine operating or a signal is not received from the ECM. The fan will also go into high-speed mode if the hydraulic hose is blocked with foreign material on the outlet side of the electronic control valve.

Other hydraulically-driven engine cooling fan systems use a thermo valve and electric solenoid instead of an electronic control valve. The thermo valve is the primary means of fan motor speed control. However, the solenoid acting under engine ECM control interrupts hydraulic fluid flow through the thermo valve, sending the fan into high speed. The purpose of this function is to protect the engine and cooling system in the event of a thermo valve failure.

Sample Test for Practice

Sample Test

Please note the letter and number in parentheses following each question. They match the task in Section 4 that discusses the relevant subject matter. You may want to refer to the overview using the cross-referencing key to help with questions posing problems for you.

1. When removing an alternator, Technician A says that you should remove the positive battery cable first. Technician B says that all of the wires on the alternator are polarized and that you need not worry about labeling them before removal. Who is right?
   A. A only
   B. B only
   C. both A and B
   D. neither A nor B   (D6)

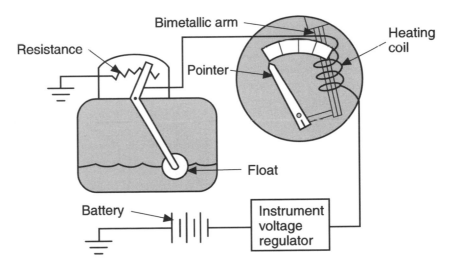

2. The fuel gauge in the figure reads lower than the actual fuel tank level suggests it should. All of the other dash gauges operate normally. Which of these is the possible cause?
   A. high resistance in the sending unit ground
   B. high resistance after the voltage regulator
   C. a short to ground between the gauge and the sending unit
   D. an open circuit in the wire from the gauge to the sending unit   (F1)

3. While performing a battery drain test, the master/key switch should be in the
   A. accessory position (Park, CL/ID).
   B. run position.
   C. night/run position.
   D. off position.   (A8)

4. A heated mirror complaint is being checked. A current draw check with an ammeter shows that the heater unit draws a varying amount of amperage. Technician A says that this may be due to the fact that the resistance of the heater decreases as it warms up. Technician B states that heater units draw variable current during normal operation. Who is right?
   A. A only
   B. B only
   C. both A and B
   D. neither A nor B   (G5)

5. While diagnosing a bus with electronic engine management and a "no-start" complaint, Technician A says you should only use a digital multimeter to check voltages on a control module circuit. Technician B says that the digital multimeter should have a 10 megohm or higher impedance. Who is right?
   A. A only
   B. B only
   C. both A and B
   D. neither A nor B   (A3)

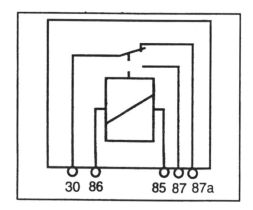

6. A 5-pin mini-relay in the figure has each pin identified with a number. What is pin #30 for?
   A. control power in
   B. control ground
   C. high amperage power in
   D. high amperage power out, normally closed   (A11)

7. Technician A says that the battery cables only need to be serviced when the starting or charging system is producing problems. Technician B says that battery corrosion only forms on the terminals during cold weather. Who is right?
   A. A only
   B. B only
   C. both A and B
   D. neither A nor B　　　　　　　　　　　　　　　　　　　　　　　　　　　　　　(C5)

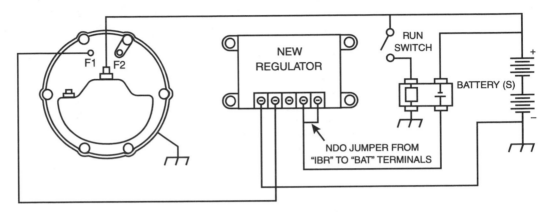

8. A bus has experienced a no-charge indicator light. Using the figure, Technician A says the problem could be a loose or corroded connection at terminal F1. Technician B says the problem could be a defective voltage regulator. Who is correct?
   A. A only
   B. B only
   C. both A and B
   D. neither A nor B　　　　　　　　　　　　　　　　　　　　　　　　　　　　　　(D5)

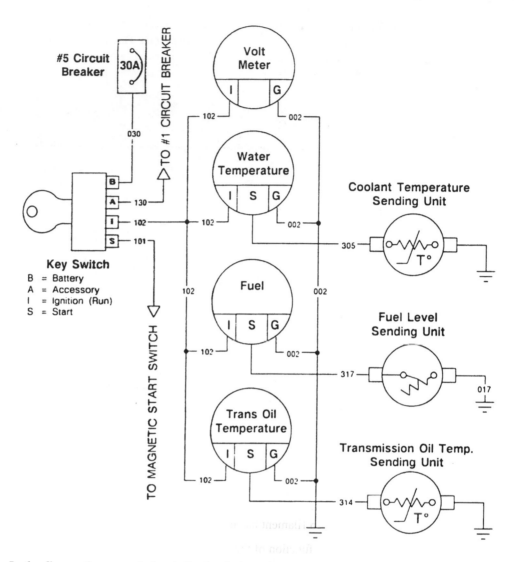

9. In the figure, the ground circuit for the fuel sender (#017) has failed. Technician A says that only the fuel gauge will be affected. Technician B states that all of the gauges will be affected because the gauges share a common ground. Who is right?
   A. A only
   B. B only
   C. both A and B
   D. neither A nor B

   (F3)

10. A maintenance-free battery is low on electrolyte. Technician A says a defective voltage regulator may cause this problem. Technician B says a loose alternator belt may cause this problem. Who is right?
    A. A only
    B. B only
    C. both A and B
    D. neither A nor B

    (D1)

11. A bus is towed to the shop with a complaint that the starter will not crank the engine. What should be checked first?
    A. ground cable connection
    B. the starter solenoid circuit
    C. the ignition switch crank circuit
    D. the battery for proper charge

    (B2)

12. Which of the following components is found in a starting circuit?
    A. a solenoid
    B. a ballast resistor
    C. a voltage regulator
    D. an alternator   (C3)

13. A bus has inoperative taillights on one side only. Technician A says to check the light switch connector for an open. Technician B uses an ohmmeter to check the continuity between the defective side and the light switch connector with the circuit under power. Who is right?
    A. A only
    B. B only
    C. both A and B
    D. neither A nor B   (E1.4)

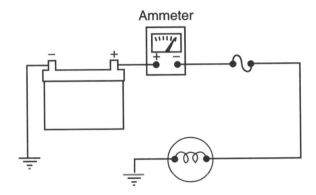

14. In the figure, the ammeter indicates that current flow through the bulb is higher than specified. The cause of this high current could be which of the following?
    A. The fuse has an open circuit.
    B. The battery voltage is low.
    C. The light bulb filament is shorted.
    D. The bulb filament has high resistance.   (A4)

15. What is the function of the park circuit in the windshield wiper system?
    A. returning the wiper blades to the start position
    B. shutting down the wiper motor in case of overheating
    C. helping to keep the wiper blades synchronized
    D. stopping the wiper motor in case of a low voltage problem   (G3)

16. When discussing a battery capacity test with the battery temperature at 70°F, Technician A says the battery discharge rate is calculated by multiplying two times the battery reserve capacity rating. Technician B says the battery is satisfactory if the voltage remains above 9.6 volts during load test at 70°F. Who is right?
    A. A only
    B. B only
    C. both A and B
    D. neither A nor B   (B1)

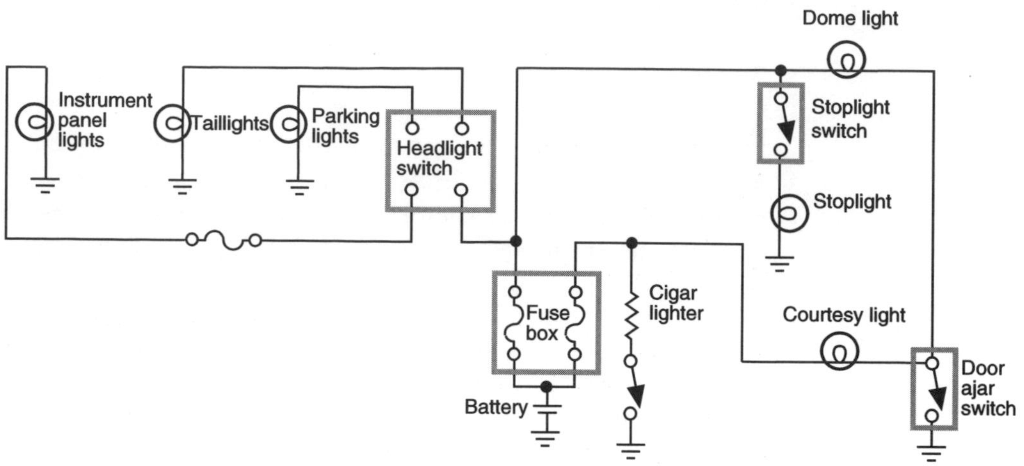

17. In the figure, the headlight switch has failed. Technician A says the parking lights cannot be affected. Technician B says that the stoplights will fail to operate. Who is right?
    A. A only
    B. B only
    C. both A and B
    D. neither A nor B   (E2.1)

18. Bus Operator No. 8823 reports that during the pre-trip inspection the emergency flashers on Bus No. 9106 are inoperative. Technician A says that the flashers should be checked for functionality or the bus operator be given another vehicle. Technician B says that this particular operator often reports defects in the hopes of getting assigned to a newer bus, and insists the operator be broken of the habit and use the original bus without inspecting it. Who is right?
    A. A only
    B. B only
    C. both A and B
    D. neither A nor B   (A1)

19. Technician A says an ammeter should be used to check for a short between circuits. Technician B says to make sure you fully charge the battery before checking a circuit for current draw. Who is right?
    A. A only
    B. B only
    C. both A and B
    D. neither A nor B   (A7)

20. A starter circuit voltage drop test checks everything EXCEPT:
    A. battery voltage.
    B. resistance in the positive battery cable.
    C. resistance in the negative battery cable.
    D. condition of the solenoid internal contacts.   (C1)

21. What is the best way to determine if a problem exists with an alternator output wire?
    A. visual inspection
    B. removing the wire and checking its resistance with an ohmmeter
    C. performing a voltage drop test with the engine off
    D. performing a voltage drop test with the alternator at maximum output   (D7)

22. The hydraulically-driven and electronically-controlled engine fan on a bus will not go into high-speed mode when the engine temperature reaches its set limit of 210°F and the engine is running. Technician A says that the electronic control valve is not getting a signal from the engine's electronic control module (ECM). Technician B says there is hydraulic hose blockage on the outlet side of the electronic control valve. Who is correct?
    A. A only
    B. B only
    C. both A and B
    D. neither A nor B   (G7)

23. The multiplex module LED indicator for the output address assigned to the driver's fan illuminates when the fan switch is turned on, but the fan does not function. All of these could be a possible cause EXCEPT
    A. a defective fan motor.
    B. a faulty wiring connection at the fan.
    C. a defective fan switch.
    D. an object lodged between the fan blades.   (A14)

24. Which of the following components could be used to progressively dim dash lights?
    A. voltage limiter
    B. rheostat
    C. transistor
    D. diode   (E1.5)

25. Warning lights and warning devices are generally activated by
    A. the vehicle master switch/key.
    B. closing of a switch or sensor.
    C. opening of a switch or sensor.
    D. the vehicle battery.   (F4)

26. Technician A says that a low battery cannot generate explosive vapors to the extent that a fully charged battery can, and therefore an explosion is far less likely. Technician B states that it is good practice to wear eye protection when jump-starting. Who is right?
    A. A only
    B. B only
    C. both A and B
    D. neither A nor B   (B6)

27. After removing a 30-amp circuit breaker from a fuse panel, a technician checks continuity across its terminals with a digital multimeter (DMM). Technician A says that the current from the ohmmeter will open the circuit breaker. Technician B says the ohmmeter should read infinity if the circuit breaker is good. Who is right?
    A. A only
    B. B only
    C. both A and B
    D. neither A nor B   (A6)

28. A bus with a traditional non-electronic flasher has a turn signal complaint. The left front light does not work and the left rear light flashes slower than normal. Technician A says the left front bulb could be defective. Technician B says there could be an open circuit between the switch and the left front bulb. Who is right?
    A. A only
    B. B only
    C. both A and B
    D. neither A nor B   (E2.3)

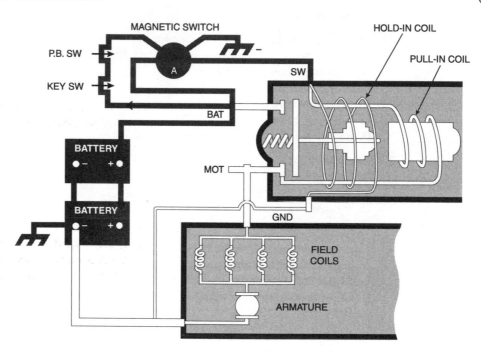

29. A bus has a no-start condition. With the key switch ("KEY SW") in the closed position (engaged) and the starter pushbutton switch ("P.B. SW") also engaged as shown in the figure, the starter engages and then immediately disengages continuously. Using the figure, Technician A says the "hold-in coil" (also referred to hold-in winding) is open. Technician B says that there is an open between the motor ("MOT") terminal and the "FIELD COILS". Who is right?
    A. A only
    B. B only
    C. both A and B
    D. neither A nor B   (C6)

30. An electric wiper motor fails to operate. Which of the following would be the LEAST likely cause?
    A. a defective wiper switch
    B. a tripped thermal overload protector
    C. a tripped circuit breaker
    D. high resistance in the motor wiring   (G2)

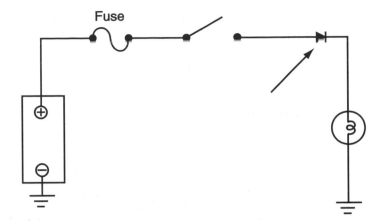

31. Technician A says when the switch is closed in the figure, the light bulb will illuminate normally. Technician B says when the switch is closed, the circuit will short, causing the fuse to burn out. Who is right?
    A. A only
    B. B only
    C. both A and B
    D. neither A nor B   (A10)

32. A bus defect condition card states that when the operator closes the rear doors, the doors open and close for several cycles before they finally stay closed. Technician A says that the rear doors are too close to each other when closed, and need to be adjusted. Technician B says that the door's sensitive edge switch needs to be adjusted. Who is correct?
    A. A only
    B. B only
    C. both A and B
    D. neither A nor B   (A16)

33. Two technicians are discussing a 12-volt charging system output test. Technician A says the vehicle accessories should be on during the test. Technician B says the charging system voltage should be limited to 17 volts. Who is right?
    A. A only
    B. B only
    C. both A and B
    D. neither A nor B   (D3)

34. Technician A says that a bus with an electronic instrument panel sources its data directly from the engine ECM. Technician B states that a data network-style electronic instrument panel receives information from the transmission ECM. Who is right?
    A. A only
    B. B only
    C. both A and B
    D. neither A nor B   (F2)

35. A problem is encountered with an interior light of a bus. Every time that a light is replaced, the fuse protecting that circuit blows. Technician A says that this could be caused by a short upstream (toward source voltage) from the light. Technician B says that this could be caused by a short downstream (toward circuit ground) from the respective light. Who is right?
    A. A only
    B. B only
    C. both A and B
    D. neither A nor B   (E1.6)

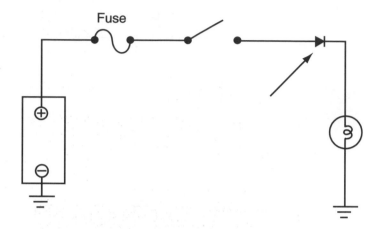

36. The arrow in the figure is pointing to a
    A. diode.
    B. resistor.
    C. capacitor.
    D. thermistor.   (A12)

37. An alternator with a 90-ampere rating produces 45 amps during an output test. A V-belt drives
    the alternator and the belt is at the specified tension. Technician A says the V-belt may be worn
    and bottomed in the pulley. Technician B says the alternator pulley may be misaligned with the
    crankshaft pulley. Who is right?
    A. A only
    B. B only
    C. both A and B
    D. neither A nor B   (D2)

38. A technician is attempting to charge a battery and yet it apparently will not accept a charge
    according to the ammeter on the charger. What is LEAST likely to be the problem?
    A. The battery is already fully charged.
    B. The battery is highly sulfated.
    C. Poor contact exists between the charging clamp and the battery post.
    D. Excessive moisture accumulation is causing surface discharge.   (B5)

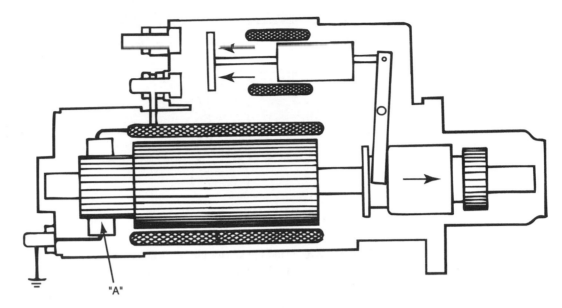

39. The item indicated by the arrow labeled "A" in the figure, if defective, will cause which of these conditions?
    A. The cranking motor turns but the engine does not.
    B. The pinion disengages slowly after starting.
    C. Unusual sounds come from the cranking motor.
    D. The engine cranks slowly but does not start. (C4)

40. A bus has an intermittent fault with its high beams only. All of these could be a possible cause EXCEPT
    A. a defective headlight dimmer switch.
    B. defective high beam filaments inside headlamps.
    C. a loose wiring harness connector.
    D. a defective headlight switch. (E1.3)

41. When welding a new bracket to a bus, one must disconnect the
    A. main power to the multiplex system.
    B. main power to the engine control module (ECM).
    C. neither A nor B
    D. both A and B (A17)

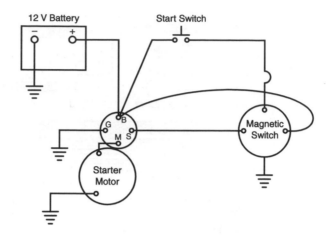

42. In the figure above, when the starter circuit is activated the magnetic switch clicks, but the starter does not operate. The technician removes the wire at point S and checks for voltage with the starting switch activated. The magnetic switch clicks and the technician measures 12 volts at the wire. The starter and solenoid assembly have been bench tested and found to be in working order. What could be the problem?
   A. The magnetic switch has a poor ground.
   B. The contacts inside the magnetic switch are badly corroded.
   C. There are faulty contacts inside the starting switch.
   D. The negative battery cable has a poor ground connection at the block.   (C2)

43. Technician A says that if a heater blower motor resistor burns out, the unit will still operate on high speed. Technician B states that if a heater blower motor resistor burns out, it will not operate in any speed position because the resistors are wired in series. Who is right?
   A. A only
   B. B only
   C. both A and B
   D. neither A nor B   (G6)

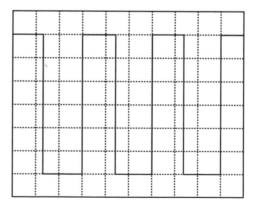

44. In the figure, when using an X1 scope probe with a voltage scale of 500mV/div, what is the peak-to-peak voltage of the signal?
   A. 15 volts
   B. 7.5 volts
   C. 3 volts
   D. 1.5 volts   (A5)

45. To perform a voltage drop test on the positive side of the charging circuit, the voltmeter connections should be at
    A. the regulator output terminal and the field terminal of the alternator.
    B. the battery ground cable and the alternator case.
    C. the output terminal of the alternator and the insulated (positive) terminal of the battery.
    D. the field terminal of the alternator and the vehicle frame.   (D4)

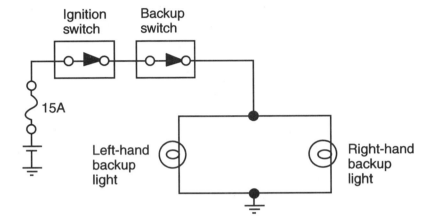

46. The right-hand backup light circuit is accidentally grounded on the switch side of the bulb in the circuit shown. Technician A says this condition may blow the backup light fuse. Technician B says the left-hand backup light may work normally while the right-hand backup light is inoperative. Who is right?
    A. A only
    B. B only
    C. both A and B
    D. neither A nor B   (E2.4)

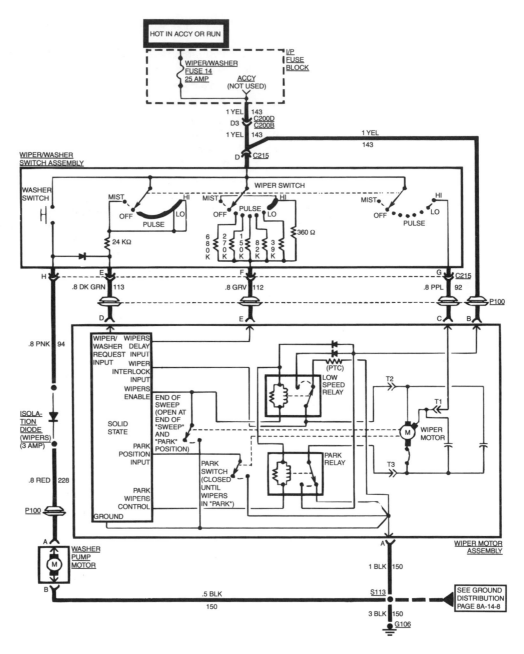

47. In the circuit shown, the windshield washer does not operate. The wiper motor operates
    normally. Technician A says the wiper/washer circuit breaker may be tripped. Technician B
    says the isolation diode may have an open circuit. Who is right?

    A. A only

    B. B only

    C. both A and B

    D. neither A nor B

                                                                                        (G4)

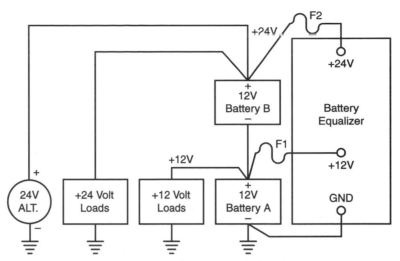

Note - Battery Banks A and B should
have the same amp-hour capacity.

48. A bus has experienced a battery discharge condition. Using the figure, Technician A says that the problem could be a bad ground connection between Battery A and the equalizer. Technician B says the problem could be a bad (open) F1 fuse. Who is correct?

 A. A only
 B. B only
 C. both A and B
 D. neither A nor B   (D8)

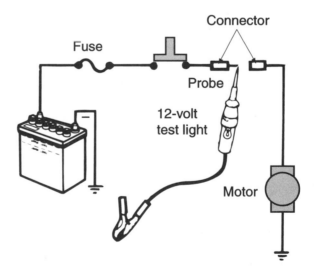

49. In the figure, the battery is fully charged and the motor is inoperative. When using a test lamp to check a circuit that does not involve an electronic module for voltage, the technician should connect the ground clip to which of these areas?

 A. the fuse
 B. the battery positive terminal
 C. the battery negative terminal
 D. the battery (hot) side of the switch   (A2)

50. The speedometer in an electronically-managed bus is not accurate. Which of the following is the LEAST likely problem?
    A. The rear axle ratio has not been correctly programmed to the engine computer.
    B. The transmission speed sensor has not been calibrated.
    C. The tire rolling radius has not been correctly programmed to the engine computer.
    D. The engine computer was not reprogrammed when new rear tires were installed.   (F5)

51. All of the following are acceptable battery and cable maintenance procedures EXCEPT
    A. removing the negative battery cable last and reinstalling first to avoid sparks.
    B. cleaning corrosion and moisture accumulation on the battery top with water and a baking soda solution.
    C. only replacing battery cable ends with proper solder or crimp on the terminals and heat shrink tubing.
    D. coating battery terminal ends with a protective grease to retard corrosion.   (B3)

52. When diagnosing a repeated flasher failure, Technician A says that a 100-millivolt voltage drop problem may be the cause. Technician B says that proper grounding of the lamp sockets may correct the problem. Who is right?
    A. A only
    B. B only
    C. both A and B
    D. neither A nor B   (E2.3)

53. An electric horn on a bus operates intermittently; which of these could be the cause?
    A. blown fuse
    B. no power to relay
    C. open in horn button circuit
    D. defective horn relay   (G1)

54. The generator failure indicator on a bus tends to intermittently come on when the bus hits a bump in the road. The generator failure indicator stays on solid when using the diagnostic module test switch while hitting the bumps. Which of the following is LEAST likely to be the cause?
    A. a defective 12-volt coil relay
    B. a loose wire
    C. a defective diagnostic module
    D. corroded connectors   (D9)

55. Technician A says that in order to read a ladder logic program, you must first start by reading condition instructions (inputs) located on the left side of the schematic. Technician B says that you can determine the inputs by solely looking at the control instructions (outputs). Who is right?
    A. A only
    B. B only
    C. both A and B
    D. neither A nor B   (A13)

56. A headlight aiming procedure is being performed. Technician A says that when headlight aiming equipment is not available, headlight aiming can be checked by projecting the high beam of each light upon a screen or chart at a distance of 25 feet ahead of the headlights. Technician B says that when aiming, the vehicle's bumper should be in direct contact with the chart or screen. Who is right?
    A. A only
    B. B only
    C. both A and B
    D. neither A nor B   (E1.2)

57. While checking the circuit breakers in a bus, the technician finds one that is tripped. The next step is to
    A. replace the circuit breaker with the next higher amperage rating.
    B. check the affected circuit for a short.
    C. check the affected circuit for an open.
    D. install a fuse of the same amperage rating as the circuit breaker.   (A9)

58. When connecting a laptop computer to a multiplex system, all of these could be viewed on the computer screen EXCEPT
    A. a directory of electrical systems controlled by the multiplex system.
    B. ladder logic rungs for the starter motor circuit.
    C. output voltage to the starter motor.
    D. status of inputs and outputs for the starter motor circuit.   (A15)

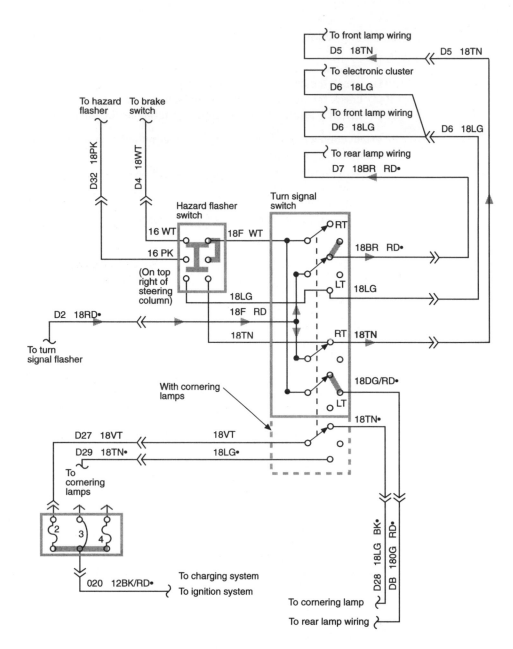

59. In the turn signal light circuit, the right rear signal light is dim, and the other lights work normally. Which of these is MOST likely the cause?
   A. high resistance in DB 180G RD from the turn signal light switch to the rear lamp wiring
   B. a short to ground in the DB 180G RD wire from the turn signal light switch to the rear lamp wiring
   C. high resistance in D7 18BR RD from the turn signal light switch to the rear lamp wiring
   D. high resistance in D2 18 RD from the turn signal flasher to the rear lamp wiring   (E2.2)

60. Technician A says that battery hold-downs should always be installed to prevent batteries from bouncing, causing possible internal damage. Technician B states that a battery box need not be cleaned when replacing a battery because the case is insulated and the battery cannot discharge because of it. Who is right?
   A. A only
   B. B only
   C. both A and B
   D. neither A nor B

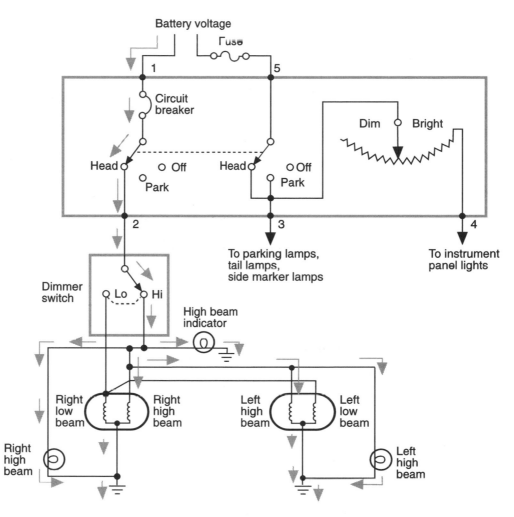

61. The left-side headlight is dim only on the high beam in the figure. The other headlights operate normally. Technician A says there may be high resistance in the left-side headlight ground. Technician B says there may be high resistance in the dimmer switch high beam contacts. Who is right?

    A. A only
    B. B only
    C. both A and B
    D. neither A nor B

                                                                                                (E1.1)

6 Additional Test Questions for Practice

Additional Test Questions

Please note the letter and number in parentheses following each question. They match the task in Section 4 that discusses the relevant subject matter. You may want to refer to the overview using the cross-referencing key to help with questions posing problems for you.

1. What is the best way to test for excessive resistance in the charging circuit?
   A. a visual inspection of the connectors
   B. a voltage drop test with the system at maximum output
   C. removing the suspect cable(s) and perform a resistance check
   D. checking battery voltage with the engine running   (D4)

2. Technician A says that a self-powered test lamp can be used to check continuity in a circuit managed by an electronic control module. Technician B states that an analog multimeter may be used to test voltages in an electronic circuit. Who is right?
   A. A only
   B. B only
   C. both A and B
   D. neither A nor B

3. When performing an alternator maximum output test, what is the most practical and safe way to make the alternator put out maximum output?
   A. place a carbon pile across the battery terminals
   B. full-field the alternator
   C. turn on all of the vehicle electrical loads
   D. install a discharged battery into the vehicle   (D3)

4. Technician A says that it is a good idea to coat the prongs and base of a new sealed beam with dielectric grease before installing the battery to prevent corrosion. Technician B says that white lithium grease can be used instead of dielectric grease. Who is right?
   A. A only
   B. B only
   C. both A and B
   D. neither A nor B   (E1.2)

5. Technician A says that a voltage drop of 500 millivolts through an electrical wire is acceptable. Technician B says that a voltage drop of 200 millivolts per connection is acceptable. Who is right?
   A. A only
   B. B only
   C. both A and B
   D. neither A nor B   (A3)

6. Technician A says that the vehicle gauges are accurate enough for diagnosing most engine problems. Technician B states that questionable readings should always be confirmed with another gauge before major repair decisions are made. Who is right?
   A. A only
   B. B only
   C. both A and B
   D. neither A nor B   (F1)

7. A driver complains of intermittent heated mirror problem. Technician A checks for a reliable power supply first. Technician B says that it is a good idea to check the mirror wiring when an intermittent problem is suspected. Who is right?
   A. A only
   B. B only
   C. both A and B
   D. neither A nor B   (G5)

8. All of these conditions would cause the starter not to crank the engine EXCEPT
   A. the battery does not connect battery power to the starter motor.
   B. the solenoid does not engage the starter drive pinion with the engine flywheel.
   C. failure of the control circuit to switch the large-current circuit.
   D. the starter drive pinion fails to disengage from the flywheel.   (C3)

9. Two technicians are discussing master/key switch off battery drain problems. Technician A states that no master/key switch off battery drain is acceptable on any bus. Technician B says battery drain can be caused by excess moisture on top of the battery. Who is correct?
   A. A only
   B. B only
   C. both A and B
   D. neither A nor B   (A8)

10. The wipers on a bus equipped with electric windshield wipers will not park. Technician A says the activation arm is broken or out of adjustment. Technician B says a defective wiper switch will cause this condition. Who is right?
    A. A only
    B. B only
    C. both A and B
    D. neither A nor B   (G3)

11. When using a voltmeter to perform a voltage drop test in a circuit, the leads should be connected in what way?
    A. to the battery terminals
    B. from the positive battery terminal to ground
    C. in series with the circuit being tested
    D. in parallel with the circuit being tested   (A3)

12. Technician A says that an overcharging alternator can cause lights that are brighter than normal. Technician B states that poor chassis grounds usually cause dim lights. Who is correct?
    A. A only
    B. B only
    C. both A and B
    D. neither A nor B   (E1.1)

13. Before replacing a battery, it is important to
    A. replace the battery cables and terminals.
    B. check the charging system.
    C. replace the starter motor.
    D. replace the alternator belt.   (B1)

14. When replacing halogen headlight bulbs, Technician A always wears gloves. Technician B says that halogen bulbs outlast conventional sealed-beam headlights. Who is right?
    A. A only
    B. B only
    C. both A and B
    D. neither A nor B   (E1.2)

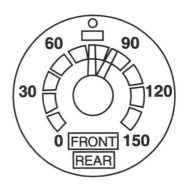

15. Technician A says that the component in the figure could be a dual ammeter. Technician B says that the component in the figure could be a dual air pressure gauge. Who is right?
    A. A only
    B. B only
    C. both A and B
    D. neither A nor B   (F1)

16. Technician A says that a starter drive pinion should not have chamfers on the drive teeth. Technician B states that if the flywheel ring gear is damaged, then the entire flywheel should be replaced. Who is correct?
    A. A only
    B. B only
    C. both A and B
    D. neither A nor B   (C4)

17. When replacing a fusible link, Technician A says you should disconnect the battery first. Technician B says you should always use the same gauge fuse link wire as the circuit being repaired. Who is right?
    A. A only
    B. B only
    C. both A and B
    D. neither A nor B   (A9)

18. When testing a battery for open circuit voltage, Technician A says that this should be done immediately after the battery comes off the charger. Technician B states that this can only be done immediately after the batteries have been charged in the vehicle. Who is right?
    A. A only
    B. B only
    C. both A and B
    D. neither A nor B   (B2)

19. The state of charge of a battery that measures a specific gravity of 1.200 at 80° F would be
    A. completely discharged.
    B. about three-fourths charged.
    C. about one-half charged.
    D. fully charged.   (B2)

20. Which of the following is NOT used when checking a circuit for continuity?
    A. an ohmmeter
    B. an ammeter
    C. a voltmeter
    D. a test lamp   (A2)

21. The "check engine" light (CEL) illuminates while a bus is being operated. Which of the following causes would not require immediate driver attention?
    A. low engine oil pressure
    B. high engine coolant temperature
    C. maintenance reminder
    D. low coolant level   (F4)

22. What is the LEAST likely cause of an engine cranking slowly?
    A. a weak battery
    B. seized pistons or bearings
    C. low resistance in the starter circuit
    D. high resistance in the starter circuit   (C1)

23. When checking open circuit battery voltage, Technician A says that a 12-volt battery is considered fully charged if a voltmeter probed across it reads anything over 12 volts. Technician B states that the battery must read at least 13.5 to 14.5 volts to be considered fully charged. Who is right?
    A. A only
    B. B only
    C. both A and B
    D. neither A nor B   (B2)

24. If an A/C compressor clutch diode fails, which of the following could result?
    A. ECM failure from voltage spikes
    B. the compressor runs backward
    C. clutch coil inoperative due to no current
    D. clutch coil failure from high current   (A10)

25. Technician A says that unwanted resistance in an electrical circuit can produce heat. Technician B states that unwanted resistance will cause reduced current flow. Who is right?
    A. A only
    B. B only
    C. both A and B
    D. neither A nor B   (A3)

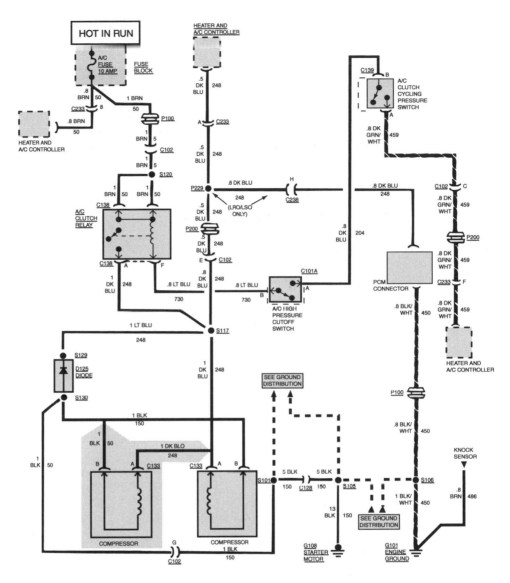

26. An electrical schematic is being examined in the figure. Technician A says that this is the heater and A/C controller circuit. Technician B says that this circuit operates with circuit #50 (1 brn) being open. Who is right?
    A. A only
    B. B only
    C. both A and B
    D. neither A nor B   (G6)

27. On a bus with electronic gauges, all of the gauge needles sweep from left to right immediately after turning on the master switch/key. Technician A says that this may indicate a fault with the electronic gauges. Technician B states that this is due to high battery voltage. Who is right?
    A. A only
    B. B only
    C. both A and B
    D. neither A nor B   (F2)

28. A bus operator reports that when the start switch is pressed, the engine does not turn over. The technician should determine if the problem is electrical or mechanical by
    A. checking the battery voltage and then turning over the engine manually before trying to activate the starting system.
    B. checking the battery voltage and immediately trying to start the engine.
    C. none of the above
    D. all of the above   (C6)

29. At 80°F, what is the correct specific gravity of electrolyte in a fully-charged battery?
    A. 1.200 to 1.220
    B. 1.220 to 1.260
    C. 1.260 to 1.280
    D. 1.280 to 1.300   (B2)

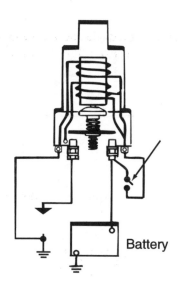

Battery

30. In the figure, to what does the arrow point?
    A. starting switch
    B. relay switch
    C. pull-in winding
    D. hold-in winding   (C2)

31. Technician A says that relays are never used by electronic control modules to control engine components. Technician B says that solenoids are found only on starter motors. Who is right?
    A. A only
    B. B only
    C. both A and B
    D. neither A nor B   (A11)

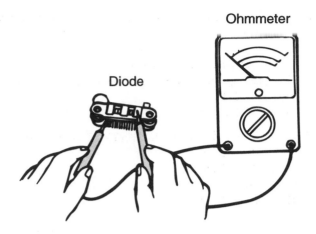

32. Technician A says a diode check is being performed in the figure. Technician B says a voltage drop test is being performed in the figure. Who is right?
    A. A only
    B. B only
    C. both A and B
    D. neither A nor B   (A10)

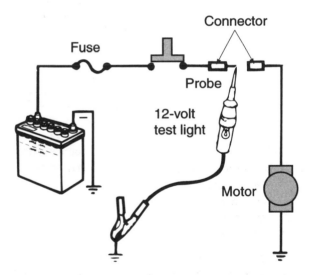

33. In the figure, Technician A says the circuit is open. Technician B says to diagnose this circuit using an ohmmeter. Who is right?
    A. A only
    B. B only
    C. both A and B
    D. neither A nor B   (A2)

34. Technician A says that when performing an alternator output test, a voltmeter should be connected in series with the alternator output terminal and the battery ground cable. Technician B says that a carbon pile should be used when performing an alternator output test. Who is right?
    A. A only
    B. B only
    C. both A and B
    D. neither A nor B   (D3)

35. When testing a circuit for voltage drop, what has to occur to obtain a reading?
    A. There must be voltage present somewhere in the circuit being tested.
    B. There must be resistance in the circuit being tested.
    C. There must be current flow in the circuit being tested.
    D. The test leads must be probed in series with the circuit being tested.   (A3)

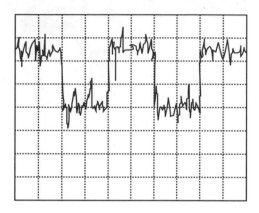

36. Using the figure, you are measuring across an input to a digital input speedometer. The input specification calls for a 5-volt peak signal. You are using a GMM with an X10 scope probe and the voltage scale is 200mV/div. What is the MOST likely problem?
    A. The probe is not calibrated.
    B. One of the signal wires is open.
    C. The speedometer is defective.
    D. There is a lot of excessive noise.   (A5)

37. Bus Operator No. 5073 reports that the rear destination sign on Bus No. 8018 occasionally does not reflect the correct route selected and it must be re-entered several times for it to display correctly. A calibrated GMM is used on the data communication line at the rear destination sign and a reduced signal is noticed when compared to other signals going to other destination signs. Technician A says that several rear destinations signs have been changed on this bus. The technician suggests a problem with defective destinations signs from the factory and insists it be changed again. Technician B says the data communication line and any associated connectors going to the rear destination sign need to be checked for integrity. Who is correct?
    A. A only
    B. B only
    C. both A and B
    D. neither A nor B   (A5)

38. Two technicians are discussing measuring resistance with an ohmmeter. Technician A says you can connect an ohmmeter in a circuit in which current is flowing. Technician B says when testing a spark plug wire with 20,000 ohms resistance, use the · 100 meter scale. Who is right?
    A. A only
    B. B only
    C. both A and B
    D. neither A nor B   (A6)

39. To establish communication with a multiplex electrical system to perform more in-depth troubleshooting of a system fault, Technician A says that with some systems this can be done with a laptop computer. Technician B says that some systems use a handheld wireless device. Who is right?
    A. A only
    B. B only
    C. both A and B
    D. neither A nor B   (A15)

40. Technician A says that all stoplight switches are air activated. Technician B states that stoplight switches route current directly to the stoplights. Who is right?
    A. A only
    B. B only
    C. both A and B
    D. neither A nor B   (E2.1)

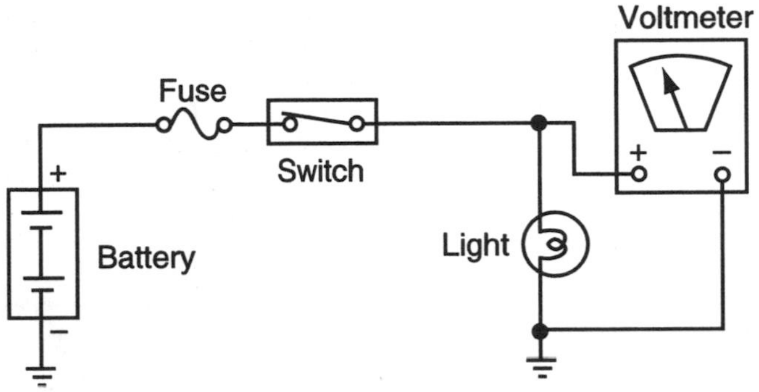

41. In the figure, the battery is fully charged and the switch is closed. With the voltmeter connected as shown, a reading of 0 volts may indicate
    A. the fuse is open.
    B. the bulb is bad.
    C. the bulb is OK.
    D. the voltmeter leads are reversed.   (A3)

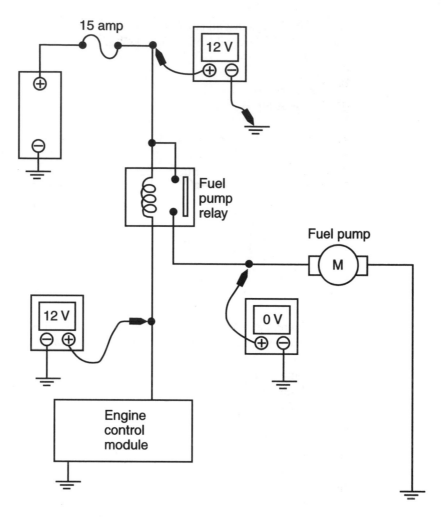

42. In the fuel pump circuit in the figure, the fuel pump is inoperative and the voltage readings shown were taken immediately after the circuit was closed. The MOST likely cause is
    A. an open 15-amp fuse.
    B. a defective engine control module.
    C. a defective fuel pump relay.
    D. a defective fuel pump.   (A11)

43. When checking cranking circuit voltage loss on the positive side of the starting circuit, the technician uses a digital multimeter on the voltage setting and performs which of these?
    A. connecting one test lead to the positive battery terminal and the other to the starter positive terminal while the starter is cranked
    B. connecting one test lead to the starter body while the starter is not turning
    C. connecting one lead to the negative side of the battery cable while the starter is turning
    D. allowing the vehicle to warm up before beginning checks   (C1)

44. While performing an open circuit voltage test, a reading of 12.4 volts is obtained. Technician A says this indicates a low battery. Technician B says the battery should be replaced. Who is right?
    A. A only
    B. B only
    C. both A and B
    D. neither A nor B   (B2)

45. When checking the power mirror and heated mirror circuit, Technician A says to replace the circuit breaker whenever the system experiences problems. Technician B says to change the fuse whenever the system experiences problems. Who is right?
    A. A only
    B. B only
    C. both A and B
    D. neither A nor B   (G5)

46. Technician A says that a multifunction switch may incorporate a windshield wiper function. Technician B says that a multifunction switch operates as a turn signal switch. Who is right?
    A. A only
    B. B only
    C. both A and B
    D. neither A nor B   (E1.3)

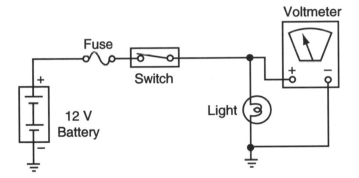

47. The battery in the figure is fully charged and the switch is closed. The voltage drop across the light bulb indicated on the voltmeter is 9 volts. Technician A says there may be a high resistance problem in the light bulb. Technician B says the circuit may be grounded between the switch and the light. Who is right?
    A. A only
    B. B only
    C. both A and B
    D. neither A nor B   (A3)

48. The dash lights on a medium-duty bus do not work. Technician A says that the fuse to the taillights could be the cause. Technician B says that the rheostat in the headlight switch could be the cause. Who is right?
    A. A only
    B. B only
    C. both A and B
    D. neither A nor B   (E1.5)

49. Which of the following is the most acceptable way to recharge a battery that is very low?
    A. fast-charge rate
    B. slow-charge rate
    C. using the vehicle's charging system
    D. adding additional electrolyte   (B5)

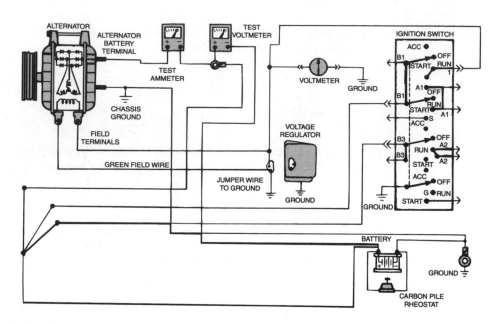

50. In the charging system in the figure, the test voltmeter reading will read
    A. charging system voltage.
    B. regulator operating voltage.
    C. charging system voltage drop.
    D. ignition switch voltage drop.

    (D4)

51. A wiper motor operates sluggishly. Technician A says that this might be due to poor brush contacts inside the motor. Technician B states that this might be caused by an open in the motor ground circuit. Who is right?
    A. A only
    B. B only
    C. both A and B
    D. neither A nor B

    (G2)

52. Ladder logic for the engine starting circuit is programmed in such a way that the starter will only be energized when the operator's starting switch is activated, the rear-start switch is placed in the front-start position, the transmission is not in gear, and the generator does not show an output signal. When reading a ladder logic diagram for the starting circuit described above, all of the statements are correct EXCEPT
    A. the starter motor will be shown on the right of the ladder logic rung.
    B. all conditions must be met before the starter will be energized.
    C. all condition instructions will be shown on the left side of the ladder logic rung.
    D. the starter motor will be energized if at least one condition is met.

    (A13)

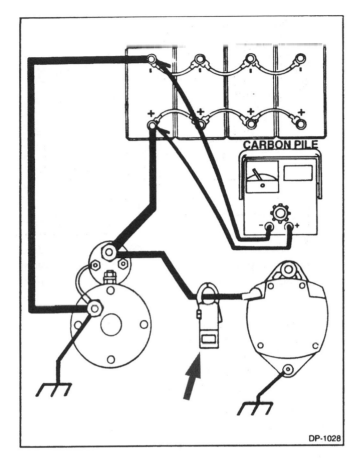

DP-1028

53. Referencing the figure, what test is being performed with the instrument indicated by the arrow (with the engine operating)?
    A. starter current draw test
    B. battery load test
    C. alternator output test
    D. parasitic battery draw test   (D3)

54. If a dash voltage limiter fails, it could cause
    A. all dash gauges to read high.
    B. the temperature gauge to read low.
    C. all dash gauges to read low.
    D. erratic operation of one or more gauges.   (F1)

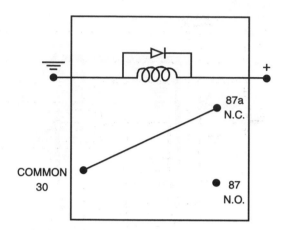

55. The figure above shows a relay with a clamping diode. Technician A says the diode makes the relay coil polarity sensitive. Technician B says polarity must only be followed if the relay is being used to control electronic devices. Who is right?
    A. A only
    B. B only
    C. both A and B
    D. neither A nor B   (A16)

56. When rebuilding alternators, Technician A always replaces the drive pulley. Technician B says that it is good practice to replace the field brushes when performing an alternator rebuild. Who is right?
    A. A only
    B. B only
    C. both A and B
    D. neither A nor B   (D6)

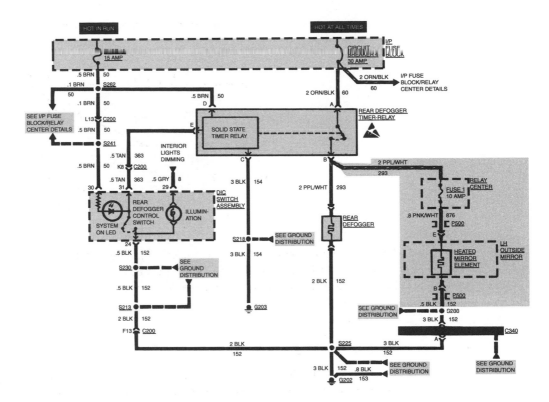

57. In the circuit shown, the heated mirror element does not function, but the rear defogger operates normally. All of the following could cause this problem EXCEPT
    A. an open circuit in the #4 circuit breaker in the fuse block.
    B. a blown #1 fuse in the relay center.
    C. an open circuit between the heated mirror element and ground.
    D. an open circuit between the timer relay and the #1 fuse.   (G5)

58. What is the LEAST likely result of a full-fielded alternator?
    A. low battery voltage due to excessive field current draw
    B. burned-out light bulbs on the vehicle
    C. a battery that boils over
    D. excessive charging system voltage   (D3)

59. Technician A says that overcharging a battery will not cause significant long-term damage. Technician B says that in hot weather, more current is needed to charge a battery. Who is right?
    A. A only
    B. B only
    C. both A and B
    D. neither A nor B   (B5)

60. A bus equipped with a multiplex electrical system has no communication or power to any of the I/O modules. Technician A says that the cause might be a faulty power supply. Technician B says that the system might be in sleep or standby mode. Who is correct?
    A. A only
    B. B only
    C. both A and B
    D. neither A nor B   (A14)

61. Technician A says an ammeter is used to test continuity. Technician B says an ammeter indicates current flow in a circuit. Who is right?
    A. A only
    B. B only
    C. both A and B
    D. neither A nor B   (A4)

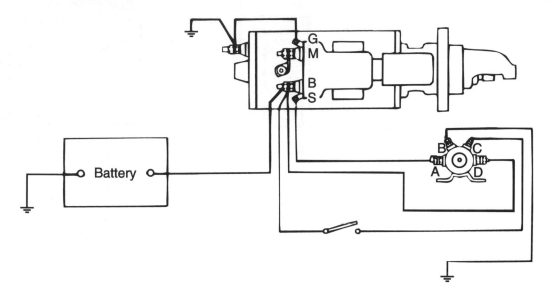

62. What component is being checked in the figure if terminals C and D are jumped?
    A. starting switch
    B. battery
    C. magnetic switch
    D. starter   (C2)

63. All the gauges are erratic in an instrument panel with thermal-electric gauges and an instrument voltage limiter. Technician A says the alternator may be at fault. Technician B says the instrument voltage limiter may be defective. Who is right?
    A. A only
    B. B only
    C. both A and B
    D. neither A nor B   (F1)

64. EMI/RFI pertains to
    A. electronic management of fuel injection.
    B. electromagnetic and radio frequency interference.
    C. electronic management and radio frequency interfaces.
    D. electronic management and radio frequency issues.   (A17)

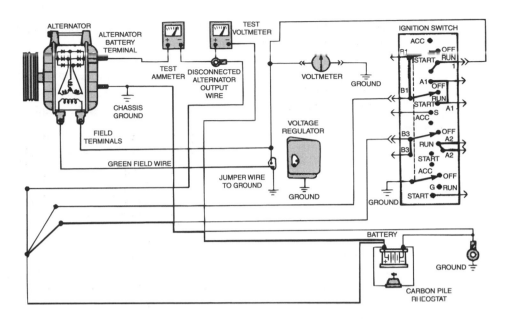

65. With an ammeter and voltmeter connected to the charging system as shown in the figure, the voltmeter indicates 2 volts and the ammeter reads 100 amps. Technician A says this condition may cause an undercharged battery. Technician B says this condition may result in a headlight flare-up during acceleration. Who is right?
    A. A only
    B. B only
    C. both A and B
    D. neither A nor B   (D4)

66. While jump-starting a vehicle with a booster battery, Technician A says the accessories should be turned on in the booster vehicle while cranking the vehicle being boosted. Technician B says the negative booster cable should be connected to the battery ground terminal(s) on the vehicle being boosted. Who is right?
    A. A only
    B. B only
    C. both A and B
    D. neither A nor B   (B6)

67. The generator failure indicator is lit. All of these could be a possible cause EXCEPT a
    A. defective generator.
    B. defective relay.
    C. defective diagnostic module.
    D. defective indicator.   (D9)

68. When mounting a battery equalizer, all of the following conditions are recommended EXCEPT
    A. providing adequate clearance for heat dissipation.
    B. providing a sealed, airtight environment.
    C. providing protection against rain and moisture.
    D. providing protection from metal coming in contact with terminals.   (D8)

69. Which of these is the LEAST likely cause of single dim headlight?
    A. corrosion on the headlight connector
    B. damaged headlamp assembly
    C. alternator output low
    D. high resistance in the wiring to the headlamp assembly   (E1.1)

70. You are working on a bus and find that a battery cable terminal end is badly corroded. All of the following are proper repair procedures EXCEPT
    A. replacing the entire cable assembly.
    B. replacing the terminal with a bolt-on end and heat shrink tubing.
    C. replacing the terminal with a crimp-on end and heat shrink tubing.
    D. replacing the terminal with a soldered end and heat shrink tubing.   (C5)

71. An alternator is overcharging. Technician A says that this can only be caused by a defective voltage regulator. Technician B states that this can be caused by excessive resistance in the charging circuit wiring. Who is right?
    A. A only
    B. B only
    C. both A and B
    D. neither A nor B   (D1)

72. A bus equipped with an electric starting motor has a slow cranking problem. All of these could be a possible cause EXCEPT
    A. corrosion or dirty battery cable connections.
    B. too small of a battery cable installed.
    C. engine with low compression.
    D. low battery voltage.   (C6)

73. A heavy-duty bus windshield wiper system is being inspected. Technician A says that air or electricity drives the wipers. Technician B says that one or two motors operate a wiper system. Who is right?
    A. A only
    B. B only
    C. both A and B
    D. neither A nor B   (G2)

74. While performing a preventative maintenance check on a mobility aid/wheelchair lift, the platform fails to retract properly into the slide channels when initially stowing the lift from being deployed. What is the MOST likely problem?
    A. the stowed/deployed limit switch
    B. the floor height proximity switch
    C. the platform chains need to be adjusted.
    D. the stow height proximity switch needs to be adjusted.   (A16)

75. An alternator output test is being performed. Technician A uses only a voltmeter connected across the battery positive terminal and negative terminal while the engine is running. Technician B says a carbon pile is not needed since the engine is already running. Who is right?
    A. A only
    B. B only
    C. both A and B
    D. neither A nor B   (D3)

76. Technician A says that a DMM can be used to test current flow directly through the meter in any bus electrical circuit. Technician B states that to test for high amperage, a current clamp will prevent damage to the meter. Who is right?
    A. A only
    B. B only
    C. both A and B
    D. neither A nor B   (A4)

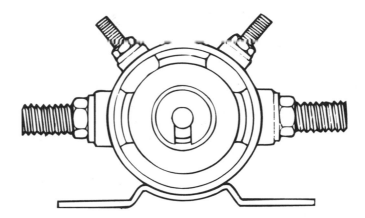

77. What is pictured in the figure?
    A. magnetic switch
    B. starting-safety switch
    C. starting switch
    D. starter solenoid   (C2)

78. A bus operator reports an intermittent problem with a mobility aid/wheelchair lift barrier only when picking up a passenger at bus stops with a high crown in the road. To correct the problem, what should the mechanic do?
    A. Replace the microswitch controlling the barrier because the switch is known to be problematic.
    B. Seek to obtain more information from the operator about the conditions under which the problem occurs.
    C. Tell the operator to have wheelchair passengers use bus stops with more level conditions.
    D. Knowing that operators are not qualified mechanically, have another technician try to duplicate the condition and together repair the problem.   (A1)

79. Shrink tubing is used to
    A. provide complete electromagnetic interference (EMI) protection.
    B. provide complete radio frequency interference (RFI) protection.
    C. improve electrical/electronic system diagnostics.
    D. insulate and protect wire connections and terminals.   (A17)

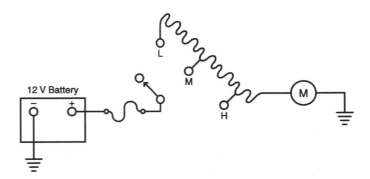

80. The blower motor in the circuit shown draws 12 amps in the high speed position. How many amps would you expect it to draw in the medium speed position?
    A. less than 12
    B. more than 12
    C. 12
    D. none   (A4)

81. A medium-duty bus with a dead battery is being jump-started. Technician A says the engine should be running on the boost vehicle before attempting to crank the dead vehicle. Technician B says the engine should be off while connecting the booster cables. Who is right?
    A. A only
    B. B only
    C. both A and B
    D. neither A nor B   (B6)

82. When cleaning or servicing the battery, cables, hold-down, and tray, you should
    A. use an air nozzle to blow off the components.
    B. use a cleaning solvent such as mineral spirits.
    C. inspect and clean the terminals if needed.
    D. use sulfuric acid to dissolve the residues.   (B3)

83. How is a starter ground circuit resistance check performed?
    A. A voltmeter is connected between the ground terminal of the battery and starter ground stud and read while the engine is being cranked.
    B. An ohmmeter is connected between the starter relay housing and the starter housing.
    C. An ohmmeter is connected between the ground side of the battery and the starter housing and read while the engine is cranking.
    D. A voltmeter is connected between the positive side of the battery and the starter solenoid while the engine is off.   (C1)

84. Technician A uses a test lamp to detect resistance. Technician B uses a jumper wire when verifying the operation of circuit breakers, relays, and lights. Who is right?
    A. A only
    B. B only
    C. both A and B
    D. neither A nor B   (A6)

85. What is the LEAST likely cause of a discharged battery?
    A. a loose alternator belt
    B. a corroded battery cable connection
    C. a defective starter solenoid
    D. a parasitic drain   (B3)

86. While testing for a possible problem with the charging system, a technician uses the diagnostic module test switch and notes that some of the indicators either dimly light or do not light at all while others are OK. Which of the following will MOST likely identify the problem?
    A. There is a problem with the ground for the diagnostic module.
    B. A shorted generator is the problem.
    C. The diagnostic module circuit breaker has tripped.
    D. The battery voltage is too low.   (D9)

87. Technician A says that undercharging could be caused by a loose drive belt. Technician B says that undercharging could be caused by undersized wiring between the alternator and the battery. Who is right?
    A. A only
    B. B only
    C. both A and B
    D. neither A nor B   (D2)

88. A starter solenoid makes a loud clicking noise when the key switch is turned to the start position, but the starter motor fails to rotate. A check of battery voltage indicates the battery is fully charged. Which of the following is MOST likely the problem?
    A. defective magnetic switch
    B. defective starting switch
    C. defective internal solenoid contacts
    D. open circuit between the magnetic switch and the starter solenoid   (C3)

89. A small automotive-type bus with a starting-safety switch is brought in for service. Technician A says that the starting-safety switch is used to prevent vehicles with automatic transmissions from being started in gear. Technician B states that starting-safety switches are often called neutral safety switches. Who is right?
    A. A only
    B. B only
    C. both A and B
    D. neither A nor B   (C2)

90. Technician A says that electric gauges can always be checked by removing the wire from the sender and grounding it to the block. Technician B states that a good way to test electric gauges for accuracy is to substitute the sender value using a variable resistance test box. Who is right?
    A. A only
    B. B only
    C. both A and B
    D. neither A nor B   (F3)

91. When checking an electric coolant temperature gauge for an erratic reading, Technician A uses an ohmmeter to check gauge resistance. Technician B places the sensing bulb in boiling water and compares the gauge reading against that of a thermometer placed in the water. Who is right?
    A. A only
    B. B only
    C. both A and B
    D. neither A nor B   (F3)

92. You are testing alternator output. Immediately after starting the engine, but before loading the battery, you find that the current output from the alternator slowly decreases the longer the engine runs. What can this mean?
    A. Alternator output is marginal; discontinue the test.
    B. The alternator drive belt is most likely slipping on the pulley.
    C. The battery is slowly recovering to capacity.
    D. The diodes in the alternator are heating up and starting to fail.   (D3)

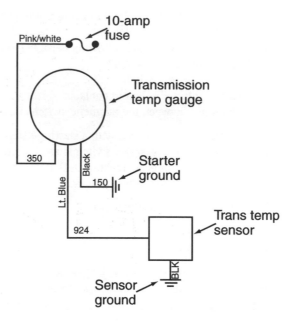

93. In the figure, what would happen if circuit 924 were open?
    A. The gauge would read approximately 10°F higher than the actual temperature.
    B. The gauge would read approximately 10°F lower than the actual temperature.
    C. The gauge would fluctuate.
    D. The gauge would be inoperative.   (F3)

94. All of the following could cause an inaccurate gauge reading EXCEPT
    A. a defective ground at the sender unit.
    B. high battery voltage.
    C. a defective sending unit.
    D. high resistance in the gauge wiring.   (F3)

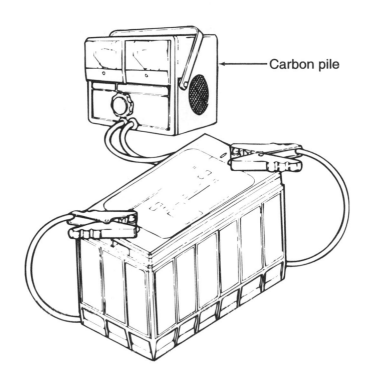

Carbon pile

95. In the figure, what test is being performed?
    A. a battery drain test
    B. a battery capacity test
    C. a battery voltage test
    D. a state of charge test   (B1)

Light "on"
indicates short

Indicate
switch

Connector

Progressively disconnect
connectors starting at the load
until light goes out

Remove load
device from circuit

96. The light bulb in the figure is inoperative. A 12-volt test light is installed in place of the fuse, the test light is on, and the bulb is off. When the connector near the bulb is disconnected, the 12-volt test light remains on. Technician A says the circuit may be shorted to ground between the fuse and the disconnected connector. Technician B says the circuit may be open between the disconnected connector and the bulb. Who is right?
    A. A only
    B. B only
    C. both A and B
    D. neither A nor B   (A7)

97. When charging a battery, you should never
    A. disconnect the negative battery cable first.
    B. charge the battery until it reads 1.265 specific gravity.
    C. reduce the fast-charging rate when specific gravity reaches 1.225.
    D. charge a frozen battery.   (B5)

98. The blower motor in a bus is running very slowly. An ammeter shows a low current flow. This could be caused by
    A. high resistance in the circuit.
    B. low resistance in the circuit.
    C. an overcharged battery.
    D. a shorted blower motor.   (A4)

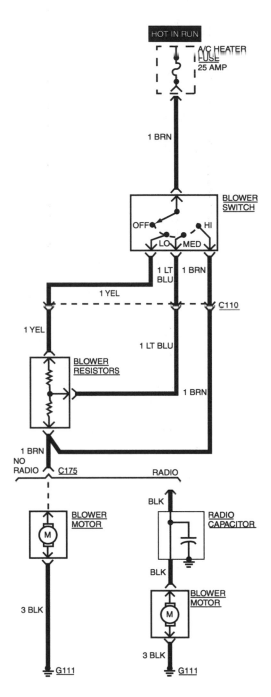

99. In the figure, when the blower resistors are removed, the blower motor will
    A. not operate.
    B. blow the system fuse.
    C. operate on high speed only.
    D. operate on low speed only.

(G6)

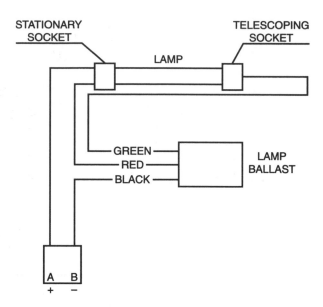

100. A bus comes in with a single overhead interior fluorescent lamp out. Using the figure, the technician performs a voltage check of the circuit at the lamp ballast on the red wire and the black wire connections, and finds a full 24-volt reading. The technician proceeds to uninstall the lamp, inspects both tabs and wiring connections in the stationary socket, and finds everything to be intact. The technician also checks the single tab and wiring connection in the telescoping socket and finds both of them to be intact. Powering the circuit down, the technician isolates the green wire from the circuit and performs a resistance check and finds the wire to have good continuity. The technician reinstalls the green wire, replaces the fluorescent lamp with a known good lamp, powers on the circuit, and still finds that there is no light. The problem is MOST likely
    A. an open red wire
    B. a bad bulb
    C. an open black wire
    D. a bad lamp ballast   (E1.6)

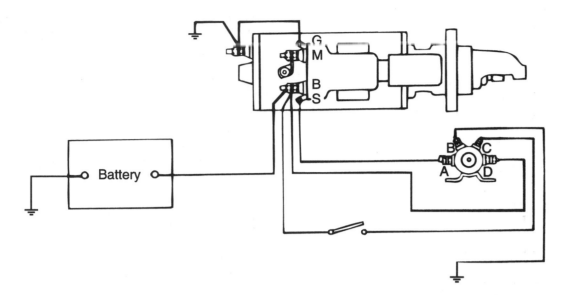

101. A technician is to perform a voltage drop test across the starter solenoid internal contacts. Using the figure, where should the voltmeter leads be placed?
    A. between the positive battery terminal and point B
    B. between the positive battery terminal and point M
    C. between points B and M
    D. between points G and ground   (C3)

102. On a bus with electronic instrumentation, coolant temperature gauge accuracy is suspect. What would be the recommended method to determine where the fault lies?
    A. swap the gauge with a like-new one
    B. replace the suspect sender unit(s)
    C. ground the sender wire at the suspect sender unit(s)
    D. connect a digital diagnostic reader and compare display to gauge readings   (F2)

103. All of these are part of the starter control circuit EXCEPT the
    A. starting switch.
    B. starter motor.
    C. starting-safety switch.
    D. magnetic switch.   (C2)

104. Technician A says that when checking an inoperative component, it should be checked downstream in the insulated side of the circuit from that component for an open. Technician B says to use an ammeter when checking for continuity in a circuit. Who is right?
    A. A only
    B. B only
    C. both A and B
    D. neither A nor B   (A7)

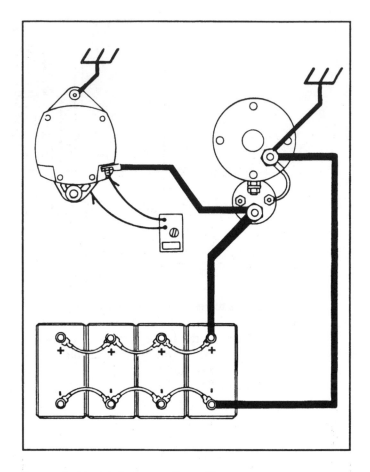

105. In the figure above, what test is being performed?
    A. voltage output test
    B. positive charging circuit cable voltage drop test
    C. charging ground circuit voltage drop test
    D. starter operating voltage test   (A3)

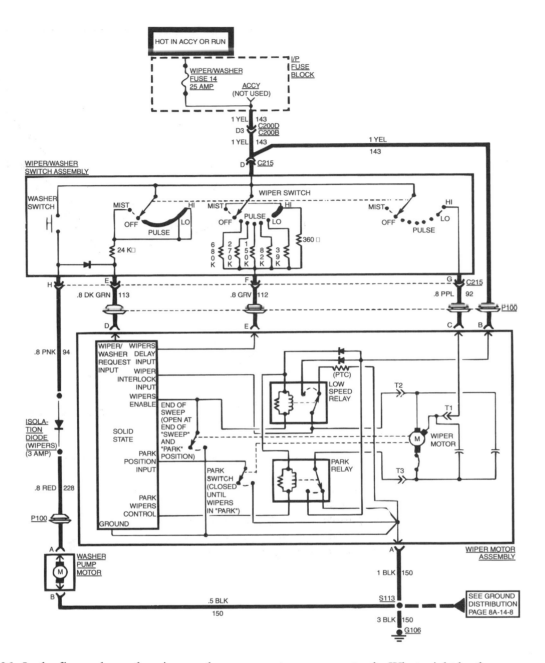

106. In the figure above, the wiper-washer pump motor runs constantly. What might be the problem?
     A. The ground side of the motor is shorted to ground.
     B. The control switch is shorted to ground.
     C. The contacts in the switch are stuck closed.
     D. The wiper washer pump relay contacts are stuck closed.   (G4)

107. Technician A says that most blower motors used in medium-duty buses use AC voltage for power. Technician B states that some blower motor circuits incorporate a relay to handle the low and medium speeds. Who is right?
     A. A only
     B. B only
     C. both A and B
     D. neither A nor B   (G6)

108. Technician A says when performing a battery load test on a 12-volt battery, a good battery should have a voltage reading below 9.6 volts while under load. Technician B says the battery should be discharged at two times its ampere-hour rating. Who is right?
    A. A only
    B. B only
    C. both A and B
    D. neither A nor B   (B1)

109. A technician is diagnosing a complaint of a windshield wiper motor that turns very slowly. A problem with high resistance in the wiring between the switch and the motor is suspected. What should be done?
    A. Disconnect the wire and measure the resistance through it.
    B. Install an ammeter in series and measure the current flow through it.
    C. Perform a voltage drop test along that length of wire.
    D. Measure the operating voltage at the motor end of the wire.   (A3)

110. Which of the following is part of a bus charging system?
    A. voltage solenoid
    B. voltage regulator
    C. voltage transducer
    D. magnetic switch   (D1)

111. The ground side of a bus charging circuit is being tested for voltage drop. Technician A says to place the voltmeter leads on the voltage regulator ground terminal and the vehicle battery. Technician B says to place the voltmeter leads on the alternator negative terminal and the battery ground terminal. Who is right?
    A. A only
    B. B only
    C. both A and B
    D. neither A nor B   (D4)

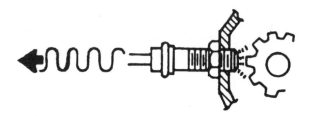

112. The type of sensor in the figure can be used for what?
    A. speed sensing
    B. pressure sensing
    C. temperature sensing
    D. level sensing   (F5)

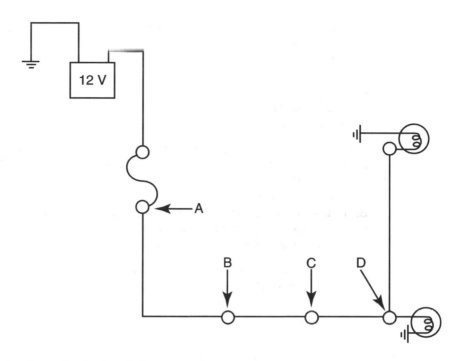

113. In the figure, the backup light circuit fuse keeps blowing. Technician A uses an ohmmeter to check the circuit between points A and B. Technician B disconnects the fuse, removes the lights, and uses an ohmmeter to check the circuit between points C and ground. Who is right?
     A. A only
     B. B only
     C. both A and B
     D. neither A nor B   (A7)

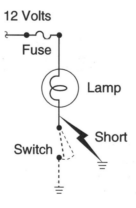

114. In the figure above, the interior operator's light circuit in a bus has a short to ground. Technician A says current flow through the lamp is higher than normal. Technician B says the light cannot be turned off. Who is right?
     A. A only
     B. B only
     C. both A and B
     D. neither A nor B   (E1.6)

115. A starter motor spins but fails to turn the engine over. Technician A says that the starter solenoid may not be activated. Technician B says that the starter drive may be defective. Who is right?
   A. A only
   B. B only
   C. both A and B
   D. neither A nor B   (C3)

116. Technician A says that a key-off current draw of 2 amps could cause the battery to discharge. Technician B says that two amps is a normal master/key switch off draw for a vehicle with an electronic control module. Who is right?
   A. A only
   B. B only
   C. both A and B
   D. neither A nor B   (A8)

117. A test of the starter control circuit may reveal problems in which area?
   A. brake interlock connections
   B. starter motor
   C. starter solenoid high-current contacts
   D. magnetic switch   (C2)

118. A bus comes in with a dim left-rear taillight. Technician A says that this may be due to a faulty switch at the operator's control panel. Technician B states that the problem may be caused by a poor ground connection at the left-rear taillight. Who is right?
   A. A only
   B. B only
   C. both A and B
   D. neither A nor B   (A7)

119. An older-type, 2-speed wiper motor is being discussed. Technician A says that 2-speed operation is enabled using separate sets of high and low speed brushes inside the motor. Technician B states that the 2-speed operation is accomplished by using an external resistor pack similar to a heater blower motor. Who is right?
   A. A only
   B. B only
   C. both A and B
   D. neither A nor B   (G2)

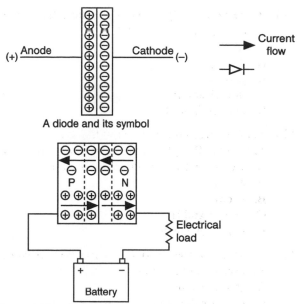

A diode and its symbol

Electrical load

Battery

Forward-biased voltage causes current flow

120. The diode in the figure is being used to
    A. allow the A/C compressor to run in one direction only.
    B. protect the A/C compressor clutch from a voltage spike.
    C. protect the circuit from a possible voltage spike.
    D. limit the current flow to the A/C compressor clutch.

(A10)

7 Appendices

Answers to the Test Questions for the Sample Test Section 5

1.	D	17.	D	33.	D	48.	C
2.	A	18.	A	34.	C	49.	C
3.	D	19.	B	35.	D	50.	B
4.	B	20.	A	36.	A	51.	A
5.	C	21.	D	37.	A	52.	D
6.	C	22.	D	38.	D	53.	D
7.	D	23.	C	39.	D	54.	C
8.	C	24.	B	40.	D	55.	A
9.	A	25.	B	41.	D	56.	A
10.	A	26.	B	42.	B	57.	B
11.	D	27.	D	43.	A	58.	C
12.	A	28.	C	44.	C	59.	C
13.	D	29.	A	45.	C	60.	A
14.	C	30.	D	46.	A	61.	D
15.	A	31.	A	47.	B		
16.	B	32.	C				

Explanations to the Answers for the Sample Test Section 5

Question #1
Answer A is wrong. The negative battery cable should always be removed first to prevent accidental shorting of your wrench to ground.
Answer B is wrong. Not all alternators have polarized connections, especially if they have been modified in the past. Be sure to label all connectors before removing.
Answer C is wrong. Neither technician is right.
Answer D is correct. Both technicians are wrong.

Question #2
Answer A is correct. High resistance in the sender ground circuit will reduce current flow, which would cause the heating coil to not bend the bi-metallic arm as far as it should.
Answer B is wrong. A high resistance after the voltage regulator should affect all the other gauges equally.
Answer C is wrong. A short to ground between the gauge and sender will cause reduced resistance and increased current flow, causing the heating coil to get hotter, bending the bi-metallic arm further than normal.
Answer D is wrong. An open circuit between the gauge and sending unit will cause zero current flow and no needle movement.

Question #3
Answer A is wrong. When the master/key switch is in the accessory (Park, CL/ID) position, there may be current being drawn through the battery to power certain loads that may be switched on. This would nullify a battery drain test.
Answer B is wrong. If the master/key switch is in the run position, there may be loads drawing current through the battery, causing an inaccurate battery drain test.
Answer C is wrong. If the ignition switch is in the night/run position, there also may be loads drawing current through the battery, causing an inaccurate battery drain test.
Answer D is correct. With the master/key switch in the off position, all normal loads are disconnected from the battery circuit. This will allow for a proper battery drain test. If a battery is found to be dead, there has to be a reason for it. Good communication between you and the operator can usually help you diagnose the problem.

Question #4
Answer A is wrong. The resistance of the heater generally increases as it warms up.
Answer B is correct. It is normal for some heated mirrors to draw more current during warm-up and then taper down in normal operation.
Answer C is wrong. Only Technician B is correct.
Answer D is wrong. One of the technicians is correct.

Question #5
Answer A is wrong. Only a DMM is an appropriate tool to use for testing voltages in an electronic circuit. However, because Technician B is also correct, answer A is wrong.
Answer B is wrong. A digital multimeter should have a minimum of 10 megohms impedance to prevent it from significantly altering a circuit's measurable parameters. However, because Technician A is also correct, answer B is wrong.
Answer C is correct. Both technicians are right.
Answer D is wrong. Both technicians are correct.

Question #6
Answer A is wrong. Control power in is pin #86.
Answer B is wrong. Control ground is pin #85.
Answer C is correct. Pin #30 is always used for high amperage power in.
Answer D is wrong. Normally closed power out is pin #87a.

Question #7
Answer A is wrong. You should service battery cables as part of regular maintenance or whenever servicing the battery.
Answer B is wrong. Battery corrosion can form on terminals in any weather. However, it affects system operation most in cold weather.
Answer C is wrong. Neither technician is correct.
Answer D is correct. Both technicians are wrong.

Question #8
Answer A is wrong. It is a good choice because excessive resistance at F1 terminal will cause low current flow resulting in a low charging rate. Yet, this is the wrong answer because both technicians are correct.
Answer B is wrong. It is a good choice because the function of the voltage regulator is to limit the alternator voltage to a pre-set value by controlling the field current. Yet, this is the wrong answer because both technicians are correct.
Answer C is correct. Both technicians are right.
Answer D is wrong. Both technicians are right.

Question #9
Answer A is correct. The ground for the fuel level gauge is independent of other grounds in the instrumentation circuit. Therefore only the fuel gauge will be affected.
Answer B is wrong. All the gauges are shown to be independently grounded.
Answer C is wrong. Only Technician A is correct.
Answer D is wrong. Only one of the technicians is wrong.

Question #10
Answer A is correct. A defective voltage regulator can cause overcharging and possible battery boilover.
Answer B is wrong. A loose alternator belt might cause undercharging but not overcharging, which is what low electrolyte level indicates.
Answer C is wrong. Only Technician A is correct.
Answer D is wrong. One of the technicians is correct.

Question #11
Answer A is wrong. The condition of the battery should always be checked first in this type of situation.
Answer B is wrong. The starter solenoid circuit would only be checked after making sure that the battery is properly charged.
Answer C is wrong. The ignition switch circuit should only be checked after verifying that the battery is fully charged.
Answer D is correct. Checking for proper battery voltage is the first step in diagnosing this type of problem.

Question #12
Answer A is correct. A solenoid is typically used on top of a starter to engage the pinion to the flywheel and make the high-current connection between the battery and starter motor.
Answer B is wrong. A ballast resistor is typically used only in some spark ignition systems.
Answer C is wrong. A voltage regulator is part of the charging system.
Answer D is wrong. An alternator is normally not involved in starting an engine.

Question #13
Answer A is wrong. An open in the switch connector would cause both taillights to be inoperative.
Answer B is wrong. An ohmmeter cannot be used to check continuity of a circuit under power.
Answer C is wrong. Both technicians are wrong.
Answer D is correct. Neither technician is right.

Question #14
Answer A is wrong. If the fuse were open, there would be no current flow at all through the circuit.
Answer B is wrong. Reduced battery voltage will cause a reduction in current flow if the resistance of the load remains the same (Ohm's law).
Answer C is correct. A shorted bulb filament will cause reduced resistance in the bulb that will cause current flow to increase if the battery voltage remains constant (Ohm's law).
Answer D is wrong. An increase in bulb resistance will cause a reduction in current flow if the battery voltage remains constant (Ohm's law).

Question #15
Answer A is correct. The function of the park switch is to return the wipers to the start position.
Answer B is wrong. The thermal overload switch shuts down the wiper motor in case of overheating.
Answer C is wrong. It is the function of the linkage to synchronize the blades (single motor systems).
Answer D is wrong. There is no low voltage protection for a wiper motor.

Question #16
Answer A is wrong. The proper discharge rate is one-half cold-cranking amps or three times the ampere-hour rating.
Answer B is correct. A battery that remains above 9.6 volts after a 15-second load test means that the battery has passed the test. This test is referred to as a load test. A load is placed on the battery to see how it functions under a demand condition.
Answer C is wrong. Only Technician B is correct.
Answer D is wrong. One of the technicians is correct.

Question #17
Answer A is wrong. The parking lights receive power from the headlight switch.
Answer B is wrong. The stoplights do not source power from the headlight switch.
Answer C is wrong. Neither technician is correct.
Answer D is correct. Both technicians are wrong.

Question #18
Answer A is correct. Only Technician A is right. All safety defects are to be corrected before the bus leaves for service. Inoperative hazard lights could result in an unsafe condition if the bus should breakdown in traffic or encounter some other trouble.
Answer B is wrong. The technician has no real way of knowing if the flashers are defective without inspecting them, and is only guessing about the operator's intent.
Answer C is wrong. Only Technician A is right.
Answer D is wrong. One of the technicians is right.

Question #19
Answer A is wrong. An ohmmeter is the best tool to use to test for shorts between circuits (with the circuit not under power). An ammeter measures current flow only; it cannot pinpoint a cross or short with another circuit.
Answer B is correct. Low battery voltage will affect current flow in a circuit.
Answer C is wrong. Only Technician B is correct.
Answer D is wrong. One of the technicians is correct.

Question #20
Answer A is correct. A starter voltage drop test does not check the condition of a battery or its state of charge, only the resistance to current flow in the various connections and cables.
Answer B is wrong. Resistance in the positive battery cables can be checked during a starter circuit voltage drop test.
Answer C is wrong. Resistance in the negative battery cables can be checked during a starter circuit voltage drop test.
Answer D is wrong. Resistance in the solenoid internal contacts can be checked during a starter circuit voltage drop test.

Question #21
Answer A is wrong. A visual inspection will not always reveal internal problems with wiring or connections.
Answer B is wrong. An ohmmeter cannot circulate enough current through a large diameter wire to effectively tell you if it has excessive resistance.
Answer C is wrong. Voltage drop tests are only meaningful with the circuit under normal full loads.
Answer D is correct. A voltage drop test under load is the best way to tell if a circuit has excessive resistance.

Question #22
Answer A is wrong. If the fan electronic control valve is not getting an electronic signal or battery voltage, the fan will automatically stay in the high-speed mode.
Answer B is wrong. If the fitting or hose is blocked on the outlet side of the electronic control module with foreign material, the fan will also stay in high-speed mode.
Answer C is wrong. Both technicians are not right.
Answer D is correct. Both technicians are wrong. In this example, the fault could be with a defective electronic control valve, causing it to stay open and allowing fluid to bypass back to the hydraulic reservoir. The spool/switching valve may also be stuck in the open position, which would also allow fluid to bypass back to the hydraulic reservoir.

Question #23
Answer A is wrong. A defective motor could be the reason for the fan not operating.
Answer B is wrong. A faulty wiring harness connection could prevent power from getting to the fan.
Answer C is correct. The illuminated indicator at the output address for the fan indicates that the fan switch has done its job by sending the appropriate command signal to the multiplex module.
Answer D is wrong. An object lodged between the fan blades could prevent the fan motor from rotating.

Question #24
Answer A is wrong. A voltage limiter is used to regulate power to instrument gauges at a constant voltage level. There is no way to adjust the voltage output with one of these.
Answer B is correct. A rheostat is a variable resistor. By turning the rheostat, the resistance is either increased or decreased, thereby changing the voltage and current flow to the bulb and altering its brightness.
Answer C is wrong. A transistor is a 3-terminal, electronic switching device.
Answer D is wrong. A diode is simply a one-way electrical check valve that allows current flow in one direction but not the other. It will not affect the resistance of the circuit in the direction of current flow.

Question #25
Answer A is wrong. The master switch/key only powers-up the circuits and is not responsible for switching individual lights on and off.
Answer B is correct. Closing a switch or sensor is usually required to complete a circuit to power-up a light or warning device.
Answer C is wrong. Opening a switch or sensor will interrupt current flow in a circuit, canceling the operation of a light or warning device.
Answer D is wrong. The vehicle battery only supplies chassis electrical power and cannot switch components on and off.

Question #26
Answer A is wrong. A low battery is MOST likely caused by excessive cranking. This is when a battery generates most of its explosive vapors.
Answer B is correct. It is always considered good practice to wear eye protection when working near batteries.
Answer C is wrong. Only Technician B is correct.
Answer D is wrong. One of the technicians is correct.

Question #27
Answer A is wrong. A DMM cannot generate anywhere near 30 amps of current when using the ohmmeter function. It will only circulate enough current through a circuit to determine its resistance.
Answer B is wrong. If a circuit breaker or fuse is good, the ohmmeter should show a very low reading (near zero), not infinity, which would indicate an open circuit.
Answer C is wrong. Neither technician is correct.
Answer D is correct. Both technicians are wrong.

Question #28
Answer A is wrong. If the front bulb fails, this will reduce current demand in the circuit, causing the flasher to blink slower than normal. Yet, this is wrong because both technicians are right.
Answer B is wrong. An open circuit to the front bulb will reduce the current demand in the circuit, making the flasher blink slower than normal and also not allow the front bulb to light. Yet, this is wrong because both technicians are correct.
Answer C is correct. Both technicians are right.
Answer D is wrong. Both technicians are correct.

Question #29
Answer A is correct. Once power is flowing through the field coils, the pull-in coil will lose its ground and will no longer be creating a magnetic field. An open hold-in coil would not be able to maintain the magnetic field needed to keep the plunger engaged. Now that the pull-in coil is at rest, the plunger will just disengage. This process will repeat itself over and over again as long as power is present at the switch terminal (SW) on the starter solenoid.
Answer B is wrong. The pull-in coil gets its ground from the "MOT" terminal through the field coils. If there is an open between the "MOT" terminal and the field coils, a circuit path could never be completed from the pull-in coil to ground and the plunger would never engage.
Answer C is wrong. Only Technician A is right.
Answer D is wrong. Technician A is right.

Question #30
Answer A is wrong. A defective wiper switch would interrupt current flow to the wiper motor and would not allow motor operation.
Answer B is wrong. A tripped thermal overload protector would open the circuit and prevent current flow to the motor.
Answer C is wrong. A tripped circuit breaker would open the circuit and not allow current flow to the motor.
Answer D is correct. High resistance in the wiring should only slow the motor down, not stop it.

Question #31
Answer A is correct. The arrow on a diode symbol points in the direction of normal current flow. The line next to the arrow point is to symbolize a blocked path to current flow. Thus, a diode allows current flow in only one direction.
Answer B is wrong. Nowhere on the positive side of the bulb is there a short to ground. A diode is simply a one-way check valve.
Answer C is wrong. Only Technician A is correct.
Answer D is wrong. One of the technicians is correct.

Question #32
Answer A is wrong. It is a good choice, but both technicians are right.
Answer B is wrong. It is a good choice, but both technicians are right.
Answer C is correct. Both conditions will cause this problem. In this case, the doors need to be properly adjusted first before the sensitive edge switch can be adjusted.
Answer D is wrong. Both technicians are right.

Question #33
Answer A is wrong. The accessories should be off to allow the alternator to direct all of its output first to the battery and then to the carbon pile.
Answer B is wrong. The charging system voltage should never exceed 15.5 volts on a 12-volt system to prevent damage to any electrical system components and may be limited to as little as 14.2 volts.
Answer C is wrong. Neither technician is right.
Answer D is correct. Both technicians are wrong.

Question #34
Answer A is wrong. Both technicians are correct.
Answer B is wrong. Both technicians are correct.
Answer C is correct. Data are received from all drivetrain systems including the engine and the transmission.
Answer D is wrong. Both technicians are correct.

Question #35
Answer A is wrong. If the short was upstream of the bulb socket, the fuse would blow regardless of whether the bulb was in place or not.
Answer B is wrong. A short to ground on the ground side of the circuit will not affect the operation of this bulb.
Answer C is wrong. Technicians A and B are both wrong.
Answer D is correct. Neither technician is correct.

Question #36
Answer A is correct. This is a symbol for a diode. The simplest device derived from using a semiconductor is a diode. It can be considered a one-way electrical check valve that allows current flow in one direction only, similar to a switch. The diode functions as both a conductor and insulator, passing current flow in one direction, called forward bias, and blocking current flow in the other direction, called reverse bias.
Answer B is wrong. This is not a symbol for a resistor. Zigzag lines usually represent these.
Answer C is wrong. A capacitor is not pictured here.
Answer D is wrong. This is not a symbol for a thermistor.

Question #37
Answer A is correct. Even though the belt is at the specified tension, it may slip because it is bottomed in the pulley. If a belt tension gauge is not available, belt tension can be determined by depressing the belt at the center of its span. As a rule, ⅜" is the approximate distance that the belt should be allowed to move.
Answer B is wrong. A misaligned pulley should not cause low alternator output because of belt slippage. It may, however, cause the belt to jump off, in which case there would be zero output.
Answer C is wrong. Only Technician A is correct.
Answer D is wrong. One of the technicians is correct.

Question #38
Answer A is wrong. A fully charged battery will not readily accept a charge.
Answer B is wrong. A highly sulfated battery will not readily accept a charge due to high internal resistance.
Answer C is wrong. If there is a poor connection between the charging clamp and the battery terminal, excess resistance will prevent the battery from being charged.
Answer D is correct. Even a heavy discharge across the top of the battery will still cause the ammeter on the charger to register some current flow through the battery.

Question #39
Answer A is wrong. If the cranking motor turns but does not start, this usually indicates a problem with the drive pinion, not the brushes.
Answer B is wrong. If the pinion disengages slowly, this usually indicates a problem with the solenoid, not the brushes.
Answer C is wrong. Noisy starter operation is usually associated with pinion problems, bent armatures, or bad bearings, not bad brushes.
Answer D is correct. Bad brushes could cause high resistance in the circuit and, therefore, slow cranking speed.

Question #40
Answer A is wrong. A defective dimmer switch can cause intermittent high beam operation.
Answer B is wrong. A loose or otherwise faulty high beam filament will cause partial or total failure of the high beam circuit.
Answer C is wrong. A loose connector in the high beam circuit can affect its operation.
Answer D is correct. A defective headlight switch should affect both low and high beam operation equally, because it powers both circuits from the same contacts.

Question #41
Answer A is wrong. It is a good choice in that power to the multiplex system needs to be disconnected, but Answer D is the correct one.
Answer B is wrong. It is also a good choice, but in this case Answer D is the correct one.
Answer C is wrong. All components with electronic modules need to be disconnected before welding on the bus.
Answer D is correct. Spikes caused by the welding process can harm electronic components on the vehicle. All electronic components need to be disconnected before doing any welding on the bus.

Question #42
Answer A is wrong. The magnetic switch would not click if it had a poor ground.
Answer B is correct. Even though the technician measured 12 volts at the wire when he activated the key, the mistake he made was to check for voltage with the circuit open. Since there was no load, 12 volts was indicated even though the contacts were corroded in the magnetic switch. He should have probed for voltage at point S with the wire connected to have the circuit under a load.
Answer C is wrong. If the starting switch contacts were defective, the magnetic switch would have not clicked.
Answer D is wrong. If the battery ground cable had a poor connection, the magnetic switch would not have clicked.

Question #43
Answer A is correct. The current for the blower motor does not have to pass through any resistors when the motor is in the high-speed position. Current from the switch bypasses the resistors.
Answer B is wrong. Even though the resistors are wired in series, the current bypasses the resistors when it is in high-speed position.
Answer C is wrong. Only one technician is correct.
Answer D is wrong. One technician is correct.

Question #44

Answer A is wrong. The vertical and horizontal lines are the divisions not the individual graticles.
Answer D is wrong. The horizontal scale is the time reference and the vertical and horizontal lines are the divisions.
Answer C is correct. The vertical and horizontal lines are the divisions and the vertical scale is voltage. Since you are using an X1 probe, the scale is read directly.
Answer D is wrong. The horizontal scale is the time reference.

Question #45

Answer A is wrong. This would be testing the voltage drop in the regulator field circuit.
Answer B is wrong. This would be testing the voltage drop on the ground side of the charging circuit, not the insulated side.
Answer C is correct. This is the correct test procedure to determine voltage drop on the insulated side of the charging circuit.
Answer D is wrong. This test would be measuring the voltage of the field circuit.

Question #46

Answer A is correct. A short to ground on the positive side of the bulb will cause increased current flow that will blow the fuse.
Answer B is wrong. Even if the fuse did not blow, the increased current demand in the circuit on the right-hand side would cause the left-hand bulb to be dim. This would happen because the two branches would no longer have equal resistance; rather, the right-hand side would have less resistance and consequent increased current flow.
Answer C is wrong. Only one of the technicians is correct.
Answer D is wrong. One of the technicians is correct.

Question #47

Answer A is wrong. A tripped circuit breaker will also cause the wiper motor to be inoperative.
Answer B is correct. If the isolation diode is open, the washer pump circuit will also be open, hence no washer pump operation.
Answer C is wrong. Only Technician B is correct.
Answer D is wrong. One of the technicians is correct.

Question #48

Answer A is wrong. It is a good choice because a loose ground will cause the equalizer not to transfer power from the 24-volt source to the 12-volt battery to maintain a voltage balance. Yet, this is the wrong answer because both technicians are correct.
Answer B is wrong. It is a good choice because a bad F1 fuse will cause the equalizer to not transfer current whenever one battery discharges at a rate different from the other. Yet, this is the wrong answer because both technicians are correct.
Answer C is correct. Both technicians are correct. The problem could be caused by both a bad ground connection between Battery A and the equalizer and a bad (open) F1 fuse.
Answer D is wrong. Both technicians are correct.

Question #49

Answer A is wrong. Connecting the test lamp ground clip at the fuse will not properly ground the test light. The fuse is on the battery positive side of this circuit.
Answer B is wrong. The battery positive terminal is also on the positive side of this circuit. The ground clip needs to be connected to a battery or chassis ground.
Answer C is correct. The battery negative terminal is a good place to clip the ground side of the test light to allow it to function properly.
Answer D is wrong. The battery positive side of the switch is also on the positive side of this circuit. The ground clip of the test light needs to be connected to a good battery or chassis ground in order to operate properly.

Question #50
Answer A is wrong. If the rear axle ratio has not been properly programmed into the engine computer, the speedometer readings will be inaccurate.
Answer B is correct. The transmission speed sensor does not need to be calibrated; it signals the engine computer output shaft speed.
Answer C is wrong. If the tire rolling radius has not been properly programmed to the engine computer, incorrect speedometer readings will result.
Answer D is wrong. Some engine management systems monitor tire wear. If new tires are installed and the engine computer is not reprogrammed, false speedometer readings will result.

Question #51
Answer A is correct. When disconnecting battery cables, always remove the ground cable first and reconnect it last to avoid sparks and a possible explosion.
Answer B is wrong. It is acceptable to clean battery tops and terminals with a baking soda and water solution.
Answer C is wrong. Battery cable terminals should only be replaced with the proper crimp type terminals along with heat shrink tubing.
Answer D is wrong. It is good practice to coat terminal ends with a protective grease to retard corrosion.

Question #52
Answer A is wrong. Even if the voltage drop was higher than the desired maximum of 100 millivolts, this would reduce current flow in the circuit, not increase it. This reduced current flow should not cause a flasher failure, although increased current flow could.
Answer B is wrong. A poor ground at the lamp socket should cause reduced current flow, not increased current flow. For this reason, it should not cause a flasher unit to fail.
Answer C is wrong. Neither technician is correct.
Answer D is correct. Both technicians are wrong.

Question #53
Answer A is wrong. A blown fuse would make the horn totally inoperative, not intermittent.
Answer B is wrong. No power to the relay would make the horn totally inoperative, not intermittent.
Answer C is wrong. An open in the horn button circuit would make the horn totally inoperative, not intermittent.
Answer D is correct. A defective horn relay could cause intermittent horn operation if its action is erratic.

Question #54
Answer A is wrong. A defective relay and its vibrating contacts can cause the generator failure indicator to intermittently come on.
Answer B is wrong. A loose wire can cause this problem.
Answer C is correct. Since using the diagnostic module test switch causes the generator failure indicator to come on solidly, the problem is elsewhere in the circuit.
Answer D is wrong. Corroded connectors can cause this problem.

Question #55
Answer A is correct. You must first understand the condition instructions or inputs located on the left side of the ladder logic rung to determine the conditions that must exist or be true before a particular output device shown on the right side of the ladder logic rung can be energized.
Answer B is wrong. Knowing the output alone is not sufficient to determine input conditions.
Answer C is wrong. Only Technician A is right.
Answer D is wrong. One of the technicians is correct.

Question #56

Answer A is correct. When headlight aiming equipment is not available, headlight aiming can be checked by projecting the upper beam of each light upon a screen or chart at a distance of 25 feet ahead of the headlights. With the headlights on high beam, the hot (brightest) spot of each headlight should be centered on the point where the corresponding vertical and horizontal lines intersect on the screen for each headlight. The headlight adjusting screws are turned in or out to adjust the headlights vertically and/or laterally to obtain a proper aim.
Answer B is wrong. Adequate distance is needed between the vehicle and the chart or screen.
Answer C is wrong. Only Technician A is correct.
Answer D is wrong. One of the technicians is correct.

Question #57

Answer A is wrong. Never replace a circuit breaker (or fuse) with one of a higher amperage rating. To do so may cause melted wiring and/or a fire.
Answer B is correct. A tripped circuit breaker usually indicates a short somewhere in the circuit causing an increase in current flow.
Answer C is wrong. If the circuit was open, current flow would stop and the circuit breaker would not have tripped in the first place.
Answer D is wrong. Installing a fuse of the same amperage rating as the circuit breaker will not solve the original problem. The fuse will merely blow.

Question #58

Answer A is wrong. A listing of all electrical systems controlled by the multiplex system can be viewed on the laptop.
Answer B is wrong. The ladder logic rungs for all multiplex circuits including the starter motor circuit can be viewed on the laptop.
Answer C is correct. Viewing the multiplex program on a laptop will not reveal any of the battery voltages present in a given circuit.
Answer D is wrong. The status of inputs and outputs for all multiplex circuits including one for the starter motor circuit can be viewed on the laptop.

Question #59

Answer A is wrong. Circuit DB 180G RD feeds the left-side rear wiring, not the right-side. This can be determined by following the wiring to the turn signal switch, where it terminates at a point marked "LT," or left turn.
Answer B is wrong. A short to ground in this wire would affect the left-side lights, not the right-side side.
Answer C is correct. Circuit D7 18BR RD feeds the right-side rear wiring. This can be determined by following the wiring to the turn signal switch, where it terminates indirectly to a point marked "RT," or right turn. High resistance in this circuit would cause a dim right rear signal light.
Answer D is wrong. High resistance in the wiring from the turn signal flasher unit to the turn signal switch would affect both left and right lamps the same.

Question #60

Answer A is correct. Batteries can be damaged internally if they are not properly secured in the battery tray.
Answer B is wrong. Batteries can self-discharge through any accumulated corrosion and moisture buildup across the top. A dirty battery tray will accelerate this process.
Answer C is wrong. Only Technician A is correct.
Answer D is wrong. One of the technicians is correct.

Question #61

Answer A is wrong. A bad ground in the left headlight would affect both high and low beam operation.

Answer B is wrong. High resistance in the dimmer switch high beam contacts would affect both right- and left-side bulbs.

Answer C is wrong. Neither technician is correct.

Answer D is correct. Both technicians are wrong.

Answers to the Test Questions for the Additional Test Questions Section 6

1.	B	31.	D	61.	B	91.	B
2.	D	32.	A	62.	A	92.	C
3.	A	33.	A	63.	B	93.	D
4.	A	34.	B	64.	B	94.	B
5.	D	35.	C	65.	A	95.	B
6.	C	36.	D	66.	D	96.	A
7.	C	37.	B	67.	D	97.	D
8.	D	38.	D	68.	B	98.	A
9.	B	39.	C	69.	C	99.	C
10.	A	40.	D	70.	B	100.	D
11.	D	41.	A	71.	B	101.	C
12.	C	42.	B	72.	C	102.	D
13.	B	43.	A	73.	C	103.	B
14.	C	44.	A	74.	D	104.	D
15.	B	45.	D	75.	D	105.	A
16.	D	46.	C	76.	B	106.	C
17.	A	47.	D	77.	A	107.	D
18.	D	48.	C	78.	B	108.	D
19.	C	49.	B	79.	D	109.	C
20.	B	50.	C	80.	A	110.	B
21.	C	51.	A	81.	C	111.	B
22.	C	52.	D	82.	C	112.	A
23.	D	53.	C	83.	A	113.	B
24.	A	54.	D	84.	B	114.	B
25.	C	55.	A	85.	C	115.	B
26.	A	56.	B	86.	A	116.	A
27.	D	57.	A	87.	C	117.	D
28.	A	58.	A	88.	C	118.	B
29.	C	59.	D	89.	C	119.	C
30.	A	60.	C	90.	B	120.	C

Explanations to the Answers for the Additional Test Questions Section 6

Question #1
Answer A is wrong. A visual inspection will usually not reveal the cause of excessive resistance.
Answer B is correct. A voltage drop test is the best way to check a circuit for excessive resistance.
Answer C is wrong. An ohmmeter cannot accurately test resistance in large diameter conductors.
Answer D is wrong. Battery voltage would have to be compared simultaneously against alternator output for the results to be meaningful.

Question #2
Answer A is wrong. A self-powered test lamp should never be used in a circuit involving an electronic module. Most electronic circuits run very low voltages. Current from a higher voltage test light may cause damage.
Answer B is wrong. An analog meter does not have the proper impedance to allow the circuit to be accurately measured.
Answer C is wrong. Neither technician is right.
Answer D is correct. Both technicians are wrong.
This test light is different from the non-powered test light in that it contains a small internal battery. It also has a bulb, probe, and a lead with an alligator clip. With its internal battery, this test light can be used to check circuits that are disconnected from the battery. A powered test light should only be used when the power to the component or circuit has been disconnected. Also, never use a powered test light on electronic circuitry. The voltage from the test light could cause damage. The powered test light is a good tool for checking continuity in a circuit as well as a ground check. To use the test light for a continuity check, place the ground clamp on the negative side of the component (ground side) and touch the positive or power side with the probe. If the component or circuit has continuity, the bulb in the test light will light. If there is an open (break in the circuit), the light will not be illuminated.

Question #3
Answer A is correct. A carbon pile tester can safely and quickly load the system to produce maximum output.
Answer B is wrong. Even though full-fielding (where applicable) is an easy way to make the alternator put out maximum output, it is a potentially dangerous test in that if the current is not dissipated somewhere, voltage in the system can rise to dangerous levels.
Answer C is wrong. All of the vehicle loads combined should not equal or exceed the alternator maximum output.
Answer D is wrong. Temporarily installing a low battery will not necessarily force the alternator to maximum output. This would depend on the condition of the battery and how low it is.

Question #4
Answer A is correct. Dielectric lubricant is the recommended lube for almost all electrical connections.
Answer B is wrong. Lithium grease is not recommended for electrical connections.
Answer C is wrong. Technician B is wrong.
Answer D is wrong. Technician A is correct.

Question #5
Answer A is wrong. Any electrical wire that drops 500 millivolts in a voltage drop test should be replaced.
Answer B is wrong. Any voltage drop test that produces a reading over 100 millivolts at any connection means that a repair is required.
Answer C is wrong. Both technicians are wrong.
Answer D is correct. Neither technician is right.

Question #6
Answer A is wrong. Vehicle gauge instrumentation is generally accurate enough for most troubleshooting procedures. However, Technician B is also correct, so answer A is wrong.
Answer B is wrong. You should never base major engine repair decisions on potentially false gauge readings without first verifying the information. However, Technician A is also correct, so answer B is wrong.
Answer C is correct. Both technicians are correct.
Answer D is wrong. Both technicians are right.

Question #7
Answer A is wrong. Checking for proper power at the source is always a good first step in any troubleshooting process. However, since Technician B is also correct, answer A is wrong.
Answer B is wrong. Faulty wiring can definitely cause intermittent operation. However, since Technician A is also correct, answer B is wrong.
Answer C is correct. Both technicians are right. Heated mirrors have two separate amperage draws: a high-amp draw during the warm-up cycle and a low-amp draw during normal operation.
Answer D is wrong. Both technicians are right.

Question #8
Answer A is wrong. If battery power was not connected to the starter motor, the starter would not rotate.
Answer B is wrong. If the solenoid does not engage the pinion with the engine flywheel, the engine will not crank.
Answer C is wrong. The control circuit is supposed to switch the large-current circuit in order to crank the engine. This is the function of the starter solenoid and the magnetic switch.
Answer D is correct. A drive pinion that fails to disengage will not cause an engine to not start. When the engine starts, the flywheel spins the pinion faster than the armature. This action releases the rollers, unlocking the pinion gear from the armature shaft. The pinion then "overruns" the armature shaft freely until being pulled out of the mesh without stressing the starter motor. Note that the overrunning clutch is moved in and out of mesh with the flywheel by linkage operated by the solenoid.

Question #9
Answer A is wrong. Buses with ECMs and other electronic modules will constantly draw some power, even with the master/key switch off.
Answer B is correct. Moisture on top of a battery can cause surface discharge between the posts.
Answer C is wrong. Only one of the technicians is correct.
Answer D is wrong. One of the technicians is correct.

Question #10
Answer A is correct. If the activation arm for the park switch is broken or out of adjustment, the wipers may not park.
Answer B is wrong. A faulty wiper switch only controls the on-off functions of the motor, not the park position.
Answer C is wrong. Only Technician A is correct.
Answer D is wrong. One of the technicians is correct.

Question #11
Answer A is wrong. Probing at the battery terminals would read battery voltage, not voltage drop in a circuit.
Answer B is wrong. This again is measuring battery voltage. The only difference is that the negative voltmeter probe is contacting ground, rather than the negative battery terminal.
Answer C is wrong. Testing a circuit in series is only done when conducting current flow tests.
Answer D is correct. Voltage drop tests are always performed by probing the leads in parallel with the component or circuit being tested.

Question #12
Answer A is wrong. An alternator that is overcharging will cause excessive system voltage that will increase the current flow through the lights causing them to be brighter than normal. However, Technician B is also correct, so answer A is wrong.
Answer B is wrong. Poor chassis grounds are usually the cause of dim lights due to rust and corrosion at mounting points. However, Technician A is also correct, so answer B is wrong.
Answer C is correct. Both technicians are correct.
Answer D is wrong. Both technicians are right.

Question #13
Answer A is wrong. Battery cables and their terminals can be reused if they are in good condition and pass a voltage drop test.
Answer B is correct. A charging system that is not operating properly can cause a battery failure. Always check for proper charging system operation when you suspect a battery failure.
Answer C is wrong. The starter motor should only be replaced if there is something wrong with it. It will have nothing to do with a battery failure.
Answer D is wrong. Belt replacement will be determined during a charging system diagnosis.

Question #14
Answer A is wrong. Halogen headlight bulbs will not tolerate oil from your skin. However, Technician B is also correct, so answer A is wrong.
Answer B is wrong. Halogen bulbs will outlast conventional sealed beams. However, Technician A is also correct, so answer B is wrong.
Answer C is correct. Both technicians are right.
Answer D is wrong. Both technicians are correct.

Question #15
Answer A is wrong. The gauge pictured cannot be an ammeter. The scaling shown does not correlate to ammeter readings.
Answer B is correct. This picture shows a dual air pressure gauge showing front and rear circuits. Air pressure gauges in air brake circuits are used to monitor air pressure in the primary and secondary circuits, and to give the driver an indication of what the application pressure is. Either individual gauges or a two-in-one gauge can be used to display the system pressure in the primary and secondary circuits. When a two-in-one gauge is used, a green or white needle typically indicates primary circuit pressure and a red needle indicates secondary circuit pressure. If a loss of pressure occurs in either of the circuits, the driver is alerted to the condition after the pressure drops below 60 psi. The alert is usually visible (gauges/warning lights) and audible (buzzer).
Answer C is wrong. Only Technician B is right.
Answer D is wrong. One of the technicians is correct.

Question #16
Answer A is wrong. A starter drive pinion is machined with a chamfer on the drive teeth to assist in meshing with the flywheel ring gear.
Answer B is wrong. A flywheel ring gear can be replaced without replacing the entire flywheel.
Answer C is wrong. Neither technician is correct.
Answer D is correct. Both technicians are wrong.

Question #17
Answer A is correct. Disconnecting the battery first prevents accidental shorting of the wiring to ground while it is being repaired.
Answer B is wrong. A fuse link should always be at least four wire gauge sizes smaller than the circuit it is protecting to allow for proper circuit protection.
Answer C is wrong. Only Technician A is correct.
Answer D is wrong. One of the technicians is correct.

Question #18

Answer A is wrong. The surface charge must be drawn off the battery if it has just come off the charger. Then the battery must be allowed to sit for 15 minutes before testing open-circuit voltage.
Answer B is wrong. A battery should never be tested for open-circuit voltage immediately after being charged or an erroneous reading will result.
Answer C is wrong. Neither technician is correct.
Answer D is correct. Both technicians are wrong.

Question #19

Answer A is wrong. A completely discharged battery will show a reading of around 1.120 or less.
Answer B is wrong. A 3/4 charged battery would show a reading of around 1.225.
Answer C is correct. A reading of 1.200 would indicate a battery that is about half-charged.
Answer D is wrong. A fully charged battery will show a reading of around 1.265 at 80°–F.

Question #20

Answer A is wrong. An ohmmeter can be used to check a circuit for continuity. The ohmmeter uses internal power to circulate a small amount of current through the circuit to check resistance and ultimately continuity.
Answer B is correct. It is not practical to use an ammeter to check for continuity. The other three choices suggest better tools for checking continuity. An ammeter is used primarily to check for the amount of current flow.
Answer C is wrong. A voltmeter can be used to probe at various points along a circuit to check that voltage exists sequentially up to the load. It can also be used to check for continuity on the ground side of a load.
Answer D is wrong. A test lamp is an easy way to check for continuity. It is used much like the voltmeter in actual operation.
The electrical term continuity refers to the circuit being continuous. For current to flow, the electrons must have a continuous path from the source voltage to the load component and back to the source. A simple vehicle circuit is made up of three parts:

1. Battery (power source)
2. Wires (conductors)
3. Load (light, motor, etc.)

Question #21

Answer A is wrong. Low engine oil pressure would require immediate action from the driver.
Answer B is wrong. High engine water temperature would also require an immediate response from the driver.
Answer C is correct. A maintenance reminder is one example of a CEL illumination that does not require the driver's immediate attention.
Answer D is wrong. Low coolant level is a potentially damaging engine condition that requires immediate attention.

Question #22

Answer A is wrong. A weak battery could cause an engine to crank slowly due to reduced voltage.
Answer B is wrong. Seized pistons or bearings could cause an engine to crank slowly because of added resistance to turning the flywheel.
Answer C is correct. Low resistance in the starter circuit will not cause an engine to crank slowly. Low resistance in the starter circuit is an ideal condition.
Answer D is wrong. High resistance in the starter circuit will cause excessive voltage drop and slower starter motor cranking speed.

Question #23
Answer A is wrong. Battery open circuit voltage must be at least 12.6 to be considered fully charged.
Answer B is wrong. This specification (13.5 to 14.5 volts) is for battery voltage while the engine is running, not for an open circuit battery test.
Answer C is wrong. Neither technician is right.
Answer D is correct. Both technicians are wrong.

Question #24
Answer A is correct. The purpose of the compressor clutch diode is to act as a spike suppression device to protect sensitive electronic components from becoming damaged due to voltage spikes.
Answer B is wrong. The direction of compressor rotation is determined by the direction the pulley is rotating.
Answer C is wrong. The clutch diode is installed in parallel with the clutch coil. Even if the diode failed, the compressor clutch will still receive voltage.
Answer D is wrong. The presence (or lack thereof) of a functioning diode will have no measurable effect on the circuit's current requirements.

Question #25
Answer A is wrong. Unwanted resistance will always produce heat in an electrical circuit; however, both technicians are right so answer A is wrong.
Answer B is wrong. Increased resistance will always cause a decrease in current flow (Ohm's law). Because both technicians are right, answer B is also wrong.
Answer C is correct. Both technicians are right.
Answer D is wrong. Both technicians are right.

Question #26
Answer A is correct. The schematic shows both heater and A/C controller components.
Answer B is wrong. An open in circuit #50 (1 brn) would prevent the A/C clutch relay from operating.
Answer C is wrong. Only Technician A is correct.
Answer D is wrong. One of the technicians is correct.

Question #27
Answer A is wrong. This gauge action occurs during an instrument self-test that verifies operation of all the gauges.
Answer B is wrong. It is not very likely that battery voltage would be too high when the master switch/key is first turned on, and even if it was the gauge action as stated is normal.
Answer C is wrong. Neither technician is correct.
Answer D is correct. Both technicians are wrong.

Question #28
Answer A is correct. When trying to determine if an engine starting problem is a mechanical or electrical one, you should first determine if sufficient battery power is present followed by determining if the engine is seized or "frozen". Should the starter be activated with a locked-up engine, considerable damage could result, including the possibility of fire.
Answer B is wrong. Although checking battery voltage is a good first step, the engine should also be turned over manually to make sure it is not locked up before attempting to engage the starter motor.
Answer C is wrong. Answer A is correct.
Answer D is wrong. Answer A is correct.

Question #29
Answer A is wrong. The range is too low for a fully charged battery.
Answer B is wrong. This range is still below the threshold for a fully charged battery.
Answer C is correct. A fully charged battery would have a reading of at least 1.265.
Answer D is wrong. These values would indicate an overcharge or possible battery damage.

Question #30

Answer A is correct The arrow is pointing to the starting switch. Current from this switch flows to the solenoid pull-in and hold-in windings (also referred to as pull-in and hold-in coils) to activate the starter.
Answer B is wrong. The figure does not show a separate magnetic or relay switch.
Answer C is wrong. The pull-in winding is located inside the solenoid, not external.
Answer D is wrong. The hold-in winding is located inside the solenoid, not external.

Question #31

Answer A is wrong. Relays are often used when a low current device (such as an ECM) needs to control a large current flow.
Answer B is wrong. Solenoids are found in many different places other than a starter. One example would be a fuel shutoff solenoid used on older buses.
Answer C is wrong. Both Technicians A and B are wrong.
Answer D is correct. Both technicians are wrong.

Question #32

Answer A is correct. The picture shows a diode being checked for continuity with an ohmmeter. In practice, the leads should then be reversed. It should show continuity in one direction, but not the other.
Answer B is wrong. In order to test voltage drop across a component, there must be current flowing in a complete circuit. Also, a voltmeter would be used for this test, not an ohmmeter.
Answer C is wrong. Only Technician A is correct.
Answer D is wrong. One of the technicians is correct.

Question #33

Answer A is correct. The picture shows the circuit open at the connector. It has been pulled apart for testing purposes.
Answer B is wrong. An ohmmeter should never be used on a live power circuit. Strange readings and possible meter damage will result.
Answer C is wrong. Only Technician A is correct.
Answer D is wrong. One of the technicians is correct.
An open is anything that causes an interruption in the flow of current in the circuit. Some examples of the causes of an open circuit are a broken wire, disconnected connector, burned-out light bulb, or broken terminals. Current cannot flow through the circuit; however, if testing the circuit, you will have source voltage up to the point of the open in the circuit. This would even include the load if the open is on the ground side of the load. For example, a marker light that is mounted to a frame might have lost its path to ground because the mounting screw of the light might not be completing the path from a ground point of the lamp to the frame on which it is mounted. In this example, source voltage is available up to the point of the grounding strap or connection of the lamp. A blown fuse is also considered an open circuit. With a fuse, though, a problem would have occurred to cause the fuse to blow, typically excessive current.

Question #34

Answer A is wrong. Voltmeters are always connected in parallel with the circuit being tested, not in series.
Answer B is correct. A carbon pile can draw current from the battery in excess of the alternator output.
Answer C is wrong. Only Technician B is correct.
Answer D is wrong. One of the technicians is correct.

Question #35
Answer A is wrong. You can have voltage present in a system without current flow. In that case, there can be no voltage drop.
Answer B is wrong. Although resistance causes voltage drops, it need not be present in order to test a circuit for a voltage drop.
Answer C is correct. There must be current flow in a circuit to cause a voltage drop.
Answer D is wrong. Tester probes must be placed in parallel to the circuit being tested, not in series.

Question #36
Answer A is wrong. A non-calibrated scope probe and will not cause this type of signal.
Answer B is wrong. An open wire will cause a loss of signal.
Answer C is wrong. A defective speedometer is unlikely to have any effect on the signal.
Answer D is correct. This is a typical display of excessive noise or interference.

Question #37
Answer A is wrong. Never change parts without first diagnosing the problem even if the same problem occurs on other buses.
Answer B is correct. A chronic problem along with a reduced signal is indicative of weak and corroded connections and data communication lines.
Answer C is wrong. Only Technician B is correct.
Answer D is wrong. Technician B is correct.

Question #38
Answer A is wrong. You should never use an ohmmeter in a circuit where current is flowing. Erroneous readings and possible meter damage will result.
Answer B is wrong. In this case the scale should be set to · 1000, not · 100. Using the · 1000 scale will give much better resolution and accuracy.
Answer C is wrong. Neither technician is correct.
Answer D is correct. Both technicians are wrong.

Question #39
Answer A is wrong. Both technicians are correct.
Answer B is wrong. Both technicians are correct.
Answer C is correct. Both technicians are correct.
Answer D is wrong. Both technicians are correct.

Question #40
Answer A is wrong. A stoplight switch is usually located on the brake pedal on smaller buses with hydraulic brakes.
Answer B is wrong. The current from a stoplight switch is most often routed through the turn signal switch to allow for proper operation of the brake lights and turn signals simultaneously.
Answer C is wrong. Neither technician is correct.
Answer D is correct. Both technicians are wrong.

Question #41
Answer A is correct. A blown or open fuse will cause an open circuit, which would give a 0-volt reading on a voltmeter.
Answer B is wrong. A defective bulb would not prevent a voltage reading to the positive side of the voltmeter nor would it affect the ground side of the voltmeter.
Answer C is wrong. If the bulb was OK, battery voltage should be present at the plus side of the voltmeter.
Answer D is wrong. If the voltmeter leads were to be reversed using a DMM, the display would simply show a negative value. If this were an analog-type meter, the needle would attempt to move to the left of zero.

Question #42

Answer A is wrong. The fuse is not open because the meter immediately after the fuse indicates 12 volts.

Answer B is correct. The voltmeter between the relay and ECM indicates 12 volts. If the ECM was functioning properly, the meter would read zero volts, because this would then be the ground side of the relay (load). Because the ECM is not providing the proper ground circuit, the meter reads battery voltage because there is no current flow. Many times relays are computer controlled. A test light is not the recommended tool to use because the light might draw more current than the circuit is designed to carry and damage the computer. In these cases, a high-impedance (10 megohm) digital voltmeter can be used to test a relay circuit.

Answer C is wrong. The relay cannot be condemned because the ECM is not allowing it to be activated.

Answer D is wrong. The fuel pump is not receiving any voltage from the relay as evidenced by the zero voltmeter reading next to the fuel pump. Therefore the pump cannot be considered faulty.

Question #43

Answer A is correct. In performing a voltage drop test on the positive side of the starting circuit, one of the test leads needs to be probed to the positive battery terminal while the starter is turning.

Answer B is wrong. The starter motor must be spinning to properly check voltage drop.

Answer C is wrong. Probing the negative battery cable checks the voltage drop on the ground side of the circuit, not the insulated side.

Answer D is wrong. It is not necessary to have the vehicle warmed up to perform a voltage drop test.

Question #44

Answer A is correct. A reading of 12.4 volts indicates a battery that is not fully charged. A reading of 12.6 volts would indicate a battery that is fully charged when tested in an open circuit state.

Answer B is wrong. Just because a battery is not fully charged does not mean that the battery needs replacement. This battery should be charged and then load tested to determine serviceability.

Answer C is wrong. Only Technician A is correct.

Answer D is wrong. One of the technicians is right.

Question #45

Answer A is wrong. A circuit breaker is designed to be replaced only if it fails.

Answer B is wrong. A fuse should only be replaced if it fails.

Answer C is wrong. Neither technician is correct.

Answer D is correct. Both technicians are wrong.

Question #46

Answer A is wrong. Technician B is also correct.

Answer B is wrong. Technician A is also correct.

Answer C is correct. Both technicians are correct. Multifunction switches may incorporate cruise control and may also serve as a turn signal switch.

Answer D is wrong. Both technicians are correct.

Question #47

Answer A is wrong. Ideally, all 12 volts from the battery should be "dropped" across the load regardless of the resistance of the bulb. A reading of 9 volts across the bulb indicates that 3 volts are being lost elsewhere in the circuit. This is the whole purpose of voltage drop testing.

Answer B is wrong. If there was a short to ground between the switch and the light, all the voltage would be going straight to ground (assuming the fuse did not blow) and there would be nothing left to be dropped across the light.

Answer C is wrong. Neither technician is correct.

Answer D is correct. Both technicians are wrong.

Question #48

Answer A is wrong. The same fuse that powers the taillights MOST likely also provides power for the dash lights. However, both technicians are correct, so answer A is wrong.

Answer B is wrong. A faulty rheostat could cause the dash lights not to come on. However, both technicians are correct, so answer B is wrong.

Answer C is correct. Both technicians are correct.

Answer D is wrong. Both technicians are correct.

Question #49

Answer A is wrong. A fast-charge rate can overheat the battery if not monitored during the process.

Answer B is correct. A slow-charge rate will prevent battery damage due to overheating from a fast-charge rate. This is also much safer because the battery will give off less gas during the process.

Answer C is wrong. A vehicle charging system will effectively fast charge a battery if it is very low. This is undesirable for the same reasons given in answer A.

Answer D is wrong. Electrolyte should never be added to a battery; only distilled water if it is low.

Question #50

Answer A is wrong. In order for the voltmeter to read charging system voltage, the negative lead of the voltmeter would need to be probing the battery negative terminal.

Answer B is wrong. The regulator in this example has been bypassed.

Answer C is correct. The voltmeter has been connected across either end of the charging system output circuit. Consequently, the voltage drop in this circuit can be measured with the system operating at rated output.

Answer D is wrong. To measure the voltage drop across the ignition switch, the voltmeter would need to be probed between a power-in and a power-out terminal on the switch itself.

Question #51

Answer A is correct. Poor brush contacts inside the motor will increase circuit resistance. This will lower current flow and cause sluggish operation.

Answer B is wrong. An open in the motor ground circuit will stop all current flow. In this case, the motor would not operate at all.

Answer C is wrong. Only Technician A is correct.

Answer D is wrong. One of the technicians is correct.

Question #52

Answer A is wrong. The out device being energized, in this case the starter motor, will be shown on the right on the ladder logic rung.

Answer B is wrong. All conditions must be met or be "true" before an output device will be energized.

Answer C is wrong. Condition instructions or inputs will be shown on the left side of the ladder logic rung.

Answer D is correct. All conditions must be met in a ladder logic circuit before the device can be energized.

Question #53

Answer A is wrong. To perform a starter current draw test, the amp clamp would have to be placed around either battery cable.

Answer B is wrong. Though the battery is being loaded with the carbon pile, the arrow in the picture specifically shows an amp clamp measuring alternator output. The battery is being loaded to force maximum output from the alternator.

Answer C is correct. The arrow shows an amp clamp being used to measure the current output through the alternator output wire.

Answer D is wrong. Parasitic battery drain tests are not done using an amp clamp. A much more sensitive current measuring device (a DMM) is needed for this test.

Question #54

Answer A is wrong. Not all gauges are affected by the voltage limiter. Some gauges are mechanical, such as the air application gauge.

Answer B is wrong. Other gauges in the panel that share the same voltage source would be similarly affected.

Answer C is wrong. Not all gauges are affected by the voltage limiter. Some gauges are mechanical, such as the air application gauge.

Answer D is correct. A malfunctioning voltage limiter would only affect those gauges which it powers.

Question #55

Answer A is correct. The diode symbol shown in the figure does in fact indicate that the relay is polarity sensitive (battery to positive, negative to ground).

Answer B is wrong. Polarity must be followed in this type of relay regardless of the application to prevent damage.

Answer C is wrong. Only Technician A is right.

Answer D is wrong. Technician A is right.

Question #56

Answer A is wrong. A pulley can be reused if it is in good condition.

Answer B is correct. Alternator brushes are subject to wear and should always be replaced during a rebuild.

Answer C is wrong. Technician A is wrong.

Answer D is wrong. Technician B is right.

Question #57

Answer A is correct. An open at circuit breaker #4 will prevent both the heated mirror and the rear defogger from operating.

Answer B is wrong. A blown #1 fuse will cause only the heated mirror to not function.

Answer C is wrong. An open between the heated mirror and ground will cause the condition described.

Answer D is wrong. An open between the timer relay and the #1 fuse will cause only the heated mirror to not function.

Question #58

Answer A is correct. A full-fielded alternator will cause maximum output from the alternator and consequent high battery voltage.

Answer B is wrong. A full-fielded alternator can cause excessive output voltage that could lead to burned out bulbs on the vehicle.

Answer C is wrong. Maximum output from an alternator due to full-fielding can cause a battery to boil over due to excessive voltage.

Answer D is wrong. A full-fielded alternator will cause excessive battery voltage because the alternator is at maximum output.

Question #59

Answer A is wrong. Overcharging will cause a battery to boil out electrolyte, and can also damage internal plates if it is severe enough.

Answer B is wrong. A battery will more readily accept a charge in warmer weather, requiring less voltage to do so.

Answer C is wrong. Both Technicians A and B are wrong.

Answer D is correct. Neither technician is correct.

Question #60
Answer A is wrong. Both technicians are correct.
Answer B is wrong. Both technicians are correct.
Answer C is correct. A faulty power supply would fail to power the modules, which in turn would result in no data communication between the multiplex components. Additionally, a system that is in sleep or standby mode would temporarily not have data communication until the system powers back up.
Answer D is wrong. Both technicians are correct.

Question #61
Answer A is wrong. While an ammeter could theoretically be used to check continuity, it would be impractical to do so. An ohmmeter is the best tool to use for this.
Answer B is correct. The purpose of an ammeter is to measure current flow.
Answer C is wrong. Only Technician B is correct.
Answer D is wrong. One of the technicians is correct.

Question #62
Answer A is correct. The starting switch is being bypassed in this test so its operation is being tested.
Answer B is wrong. Although the battery is being tested here, it is not the objective of this test.
Answer C is wrong. When terminals C and D are jumped, the magnetic switch is activated independent of the starting switch. To test the starting switch should cause it to engage.
Answer D is wrong. Although the starter is activated, it is not the objective of this test.

Question #63
Answer A is wrong. The instrument voltage limiter maintains a constant voltage value to the gauges regardless of battery voltage.
Answer B is correct. A defective instrument voltage limiter will affect all the dash gauges.
Answer C is wrong. Only Technician B is correct.
Answer D is wrong. One of the technicians is correct.

Question #64
Answer A is wrong. EMI/RFI pertains to electromagnetic interference (EMI) and radio frequency interference (RFI).
Answer B is correct. It represents the correct terminology.
Answer C is wrong. Answer B is correct.
Answer D is wrong. Answer B is correct.

Question #65
Answer A is correct. The technician is performing a voltage drop test on the insulated (positive) side of the charging circuit. A 2-volt drop exceeds specifications, and this will cause an undercharged battery.
Answer B is wrong. An excessive voltage drop will cause low voltage and dim lights, not high voltage and headlight flare-up.
Answer C is wrong. Only Technician A is correct.
Answer D is wrong. One of the technicians is correct.

Question #66
Answer A is wrong. All unnecessary loads in both vehicles should be turned off to allow the maximum amount of current to flow to the battery and the starter.
Answer B is wrong. It is best to make the last ground connection away from the batteries.
Answer C is wrong. Neither technician is correct.
Answer D is correct. The negative booster cable should be connected to the frame/chassis ground on the vehicle being boosted. When jump-starting a vehicle, the last connection made will cause a spark to occur. Making the last connection to the frame/chassis ground ensures that the spark will be as far away from the batteries as possible.

Question #67
Answer A is wrong. A defective generator is what normally causes the indicator to light.
Answer B is wrong. A defective relay can cause the indicator to light.
Answer C is wrong. A defective diagnostic module can cause the indicator to light.
Answer D is correct. Indicators will burn out or open up, causing them not to illuminate.

Question #68
Answer A is wrong. Adequate clearance is required to dissipate heat normally generated from the equalizer.
Answer B is correct. Although equalizers need protection, an airtight compartment would not allow heat generated from the equalizer to be dissipated.
Answer C is wrong. Equalizers need protection from rain and moisture.
Answer D is wrong. Equalizer terminals contain battery voltage and need to be protected from possible shorts.

Question #69
Answer A is wrong. Corrosion on the connector will create a voltage drop which will cause a dim headlight.
Answer B is wrong. A damaged headlight assembly can cause increased resistance at the connections or bulb filament, causing a voltage drop and dim lights.
Answer C is correct. Low alternator output will cause both lights to be dim.
Answer D is wrong. High resistance in the wiring leading to the bulb will cause a voltage drop and therefore decreased light output.

Question #70
Answer A is wrong. Replacing the entire battery cable is an acceptable method of repair. It may be the fastest and most economical way depending on shop preferences.
Answer B is correct. You should never replace a battery cable end with an aftermarket type bolt-on end. These have poor wire contact characteristics and the entire joint is exposed and subject to corrosion.
Answer C is wrong. Replacing a battery cable terminal end with a crimp on terminal and heat shrink tubing is considered an acceptable method of repair.
Answer D is wrong. Replacing a battery cable terminal end with a soldered-on terminal and heat shrink tubing is considered an acceptable method of repair.

Question #71
Answer A is wrong. While a defective voltage regulator is the MOST likely cause, it can also be caused by a bad sense diode on systems with external regulators.
Answer B is correct. Excessive resistance in the battery feed circuit to the voltage regulator will cause the regulator to sense that the system voltage is low and cause the alternator to charge at a higher rate. This higher charge rate will in turn overcharge the batteries.
Answer C is wrong. Only Technician B is correct.
Answer D is wrong. One of the technicians is correct.

Question #72
Answer A is wrong. Corrosion or dirty cable connections would create a voltage drop when a load is placed on the circuit, which would result in a slow cranking condition.
Answer B is wrong. A battery cable that is too small for the amperage load will create a resistance when the load is placed on the circuit, resulting in a voltage drop that creates a slow cranking condition.
Answer C is correct. An engine with low compression will create a fast cranking condition, not a slow one.
Answer D is wrong. Low battery voltage will cause a slow cranking condition.

Question #73
Answer A is wrong. Wiper systems can either be air or electric operated. However, Technician B is also correct, so answer A is wrong.
Answer B is wrong. A wiper system can use one or two motors. However, Technician A is also correct, so answer B is wrong.
Answer C is correct. Both technicians are correct.
Answer D is wrong. Neither technician is wrong.

Question #74
Answer A is wrong. The stowed/deployed limit switch signals the ECU when the lift is in the stowed/deployed position.
Answer B is wrong. The floor height proximity switch signals the ECU when the lift is at coach floor height.
Answer C is wrong. A misalignment of the chains would cause the platform to bind.
Answer D is correct. The stow height proximity switch determines the correct height where the platform and slide channels properly match.

Question #75
Answer A is wrong. An ammeter is also needed to determine the current output of the alternator.
Answer B is wrong. A carbon pile is needed to load the system to force maximum alternator output.
Answer C is wrong. Both technicians are wrong.
Answer D is correct. Neither technician is correct.

Question #76
Answer A is wrong. Most DMMs have a 10–20 amp limit when making current tests directly through the meter.
Answer B is correct. Most DMMs are not capable of measuring much more than 10–20 amps directly through a meter without blowing a fuse. For this reason an amp clamp should be used on high current circuits.
Answer C is wrong. Only Technician B is correct.
Answer D is wrong. One of the technicians is right.

Question #77
Answer A is correct. This figure shows a magnetic switch.
Answer B is wrong. This figure does not represent a starting-safety switch.
Answer C is wrong. This figure does not represent a starting switch.
Answer D is wrong. This figure does not represent a starter solenoid.

Question #78
Answer A is wrong. Parts should never be changed unless first diagnosing the cause of the problem.
Answer B is correct. Intermittent problems are best solved when additional information can be obtained from the bus operator who has first-hand knowledge of the conditions under which the problem occurs.
Answer C is wrong. Passengers should never be inconvenienced because technicians are unwilling to diagnose a mechanical problem.
Answer D is wrong. It is a good choice because two technicians should have the ability to repair the fault, but the first course of action concerning intermittent problems is to try to obtain additional information from the operator who has first-hand knowledge of the problem.

Question #79

Answer A is wrong. Shrink tubing does not provide complete EMI protection.

Answer B is wrong. Shrink tubing does not provide complete RFI protection.

Answer C is wrong. Shrink tubing has nothing to do with diagnostics.

Answer D is correct. When heat is applied, shrink tubing retracts or "shrinks" tightly around a wire connection or terminal to provide insulation and protection.

Question #80

Answer A is correct. A resistance has been added in series to the motor circuit. Ohm's law states that the current flow will decrease if the voltage stays the same and resistance is increased.

Answer B is wrong. Because a resistance has been added to the circuit, current flow will decrease in this circuit, not increase.

Answer C is wrong. Ohm's law states that the current in the circuit will decrease not stay the same.

Answer D is wrong. The circuit has not been broken (opened). Therefore, there will be some current flow.

Question #81

Answer A is wrong. By starting the booster vehicle, the charging system will produce more voltage to the battery than would be attainable if the engine were left off. This will aid in the starting process. However, since Technician B is also correct, answer A is wrong.

Answer B is wrong. The engine should be off when connecting the cables to minimize the potential for a big spark and possible explosion. However, since Technician A is also correct, answer B is wrong.

Answer C is correct. Both technicians are correct.

Answer D is wrong. Both technicians are right.

Question #82

Answer A is wrong. You should never use compressed air to blow possible battery acid around.

Answer B is wrong. Mineral spirits are not used to clean a battery.

Answer C is correct. You should always inspect and clean battery cable terminals if needed when servicing a battery to scrape away any accumulated corrosion.

Answer D is wrong. A baking soda solution should be used to dissolve battery residue, not sulfuric acid.

Question #83

Answer A is correct. When you connect a voltmeter between the battery ground terminal and the starter ground stud, a voltage drop test of the ground side of the starting circuit is performed. This checks the resistance of the ground side of the circuit.

Answer B is wrong. An ohmmeter cannot provide enough current through the starting circuit to determine excessive resistance. This can only be done with a voltage drop test.

Answer C is wrong. An ohmmeter should never be connected to a live circuit. Possible meter damage may result.

Answer D is wrong. In order to conduct a voltage drop test, the starter must be cranking. Also, testing the positive side of the battery will not verify ground circuit performance.

Question #84
Answer A is wrong. Test lamps cannot measure resistance, only ohmmeters can.
Answer B is correct. Jumper wires can be used to temporarily supply power to circuit breakers, relays, and lights to check for proper operation. A jumper wire is often overlooked as a piece of test equipment. Many times it is used as an extension of a circuit. Its value, though, comes when it is used to isolate a component, bypassing the circuit. The jumper wire has an alligator clip on each end. It should be fabricated with a fuse or circuit breaker to prevent damage to the circuit or component. Connect one end of the jumper wire to the battery and the other end to the feed side of the component to check the operation of that component. If the component now works, you should look for a problem in the circuit. Caution: Never use a jumper wire to bypass a fuse. Also, never connect a jumper wire across a battery. This will cause the wire to burn and could cause the battery to explode.
Answer C is wrong. Only Technician B is correct.
Answer D is wrong. Technician B is correct.

Question #85
Answer A is wrong. A loose alternator belt can cause belt slippage, alternator undercharging, and, consequently, a discharged battery.
Answer B is wrong. A corroded battery cable connection can cause increased resistance. The resulting voltage drop at this point could cause an undercharged battery because not all of the available alternator voltage can reach the battery to properly charge it.
Answer C is correct. A bad starter solenoid will have no direct effect on a discharged battery.
Answer D is wrong. A parasitic drain on the battery can cause a battery to discharge over a period of time.

Question #86
Answer A is correct. A grounding problem with the diagnostic module will cause the current to back-feed into other circuits, finding other ways to ground.
Answer B is wrong. The relays in the charging system circuit monitor will isolate the generator from the diagnostic module.
Answer C is wrong. Most, if any, of the indicators would not light at all when the diagnostic module test switch is activated.
Answer D is wrong. All of the indicators would dimly light or be affected.

Question #87
Answer A is wrong. A loose drive belt can cause the pulley to slip with resultant undercharging. However, Technician B is also correct, so answer A is wrong.
Answer B is wrong. Undersized wiring between the battery and the alternator can result in excessive voltage drop that can result in an undercharged battery. However, Technician A is also correct, so answer B is wrong.
Answer C is correct. Both technicians are right.
Answer D is wrong. Both technicians are correct.

Question #88
Answer A is wrong. A defective magnetic switch would not power the starter solenoid causing it to click.
Answer B is wrong. A defective starting switch will not power the magnetic switch, which in turn will not power the starter solenoid.
Answer C is correct. Defective internal solenoid contacts will MOST likely prevent starter motor operation, but not prevent the solenoid plunger from moving into the engaged position causing the clicking noise.
Answer D is wrong. An open circuit between the starter solenoid and the magnetic switch would prevent solenoid operation altogether, therefore it would not click.

Question #89
Answer A is wrong. The starting safety switch is used to prevent an engine from starting with the transmission in gear. However, Technician B is also correct, so answer A is wrong.
Answer B is wrong. Starting-safety switches can also be called neutral-safety switches. However, Technician A is also correct, so answer B is wrong.
Answer C is correct. Both technicians are correct.
Answer D is wrong. Both technicians are right.

Question #90
Answer A is wrong. While some gauges can be checked this way, doing so can damage a gauge. Consult the vehicle repair manual before using this procedure.
Answer B is correct. A variable resistance test box is a good way to substitute a resistance value to the gauge sent by a suspect sender unit.
Answer C is wrong. Only Technician B is correct.
Answer D is wrong. Only one of the technicians is wrong.

Question #91
Answer A is wrong. Checking gauge resistance is not an accurate way to test gauge performance.
Answer B is correct. This is an accurate way to test a standard temperature gauge.
Answer C is wrong. Only Technician B is correct.
Answer D is wrong. One of the technicians is right.

Question #92
Answer A is wrong. This action is normal. The battery is being replenished quickly, and so the charging current going into it will steadily decrease as time goes on.
Answer B is wrong. A slipping drive belt should produce a consistent current output from the alternator.
Answer C is correct. This is a normal occurrence as a battery is recharged.
Answer D is wrong. Diodes do not fail gradually. They either work or they do not.

Question #93
Answer A is wrong. The gauge should only read high if the sender wire was shorted to ground, not open.
Answer B is wrong. If circuit 924 was open, the gauge should not read at all.
Answer C is wrong. An open in circuit 924 should cause no action at all, not fluctuations.
Answer D is correct. An open in the sender circuit should cause the gauge to be inoperative because no current can flow through the circuit.

Question #94
Answer A is wrong. A faulty ground at the sender unit could alter the circuit's overall resistance and throw off the gauge reading.
Answer B is correct. High battery voltage is compensated for by the IVR (instrument voltage regulator). Most magnetic gauges are not affected by varying voltage levels.
Answer C is wrong. A defective sending unit would result in an inaccurate reading.
Answer D is wrong. Excessive resistance in the wiring will alter the circuit's overall resistance, throwing off the gauge readings.

Question #95
Answer A is wrong. A battery parasitic drain test is performed with a voltmeter, not a carbon pile.
Answer B is correct. The figure shows a battery being prepared for a capacity test using a carbon pile to provide the load. The battery discharge rate for a capacity test is usually one-half of the cold-cranking rating. The battery is discharged at the proper rate for 15 seconds to eliminate the surface charge. Then, after about 1 to 2 minutes, the test is repeated. The battery voltage must remain above 9.6V with the battery temperature at 70°F or above.
Answer C is wrong. A battery voltage test is done with a DMM, not a carbon pile.
Answer D is wrong. The state of charge test is done with a DMM or a hydrometer, not a carbon pile.

Question #96

Answer A is correct. If the test light remains on even after the connector is unplugged, this must mean that there is a short to ground somewhere between the test light and the unplugged connector. If there was no short to ground, the test lamp could not glow.

Answer B is wrong. If the circuit were open, there would be no way for current to flow through it and cause the test lamp to glow.

Answer C is wrong. Only Technician A is correct.

Answer D is wrong. One of the technicians is correct.

Question #97

Answer A is wrong. The negative battery cable should always be disconnected first when charging a battery.

Answer B is wrong. You should consider a battery to be fully charged when the specific gravity reaches 1.265.

Answer C is wrong. You should reduce the fast-charging rate when the specific gravity reaches 1.225.

Answer D is correct. A frozen battery should be brought to room temperature before charging to prevent damage or explosion.

Question #98

Answer A is correct. Increased resistance in a circuit with constant battery voltage will cause reduced current flow (Ohm's law).

Answer B is wrong. A decrease in the resistance of a circuit will cause an increase in current flow with constant battery voltage applied (Ohm's law).

Answer C is wrong. An increase in battery voltage will cause an increase in current flow if the resistance of the load remains constant (Ohm's law).

Answer D is wrong. A short in the blower motor will cause reduced resistance, which will cause an increase in current flow with a constant battery voltage applied (Ohm's law).

Question #99

Answer A is wrong. The blower motor will operate on the high-speed position because the resistors arc bypassed.

Answer B is wrong. Removing the resistors will create an open circuit in the low and medium speeds. Since there would be no current flow, the fuse would not blow.

Answer C is correct. With the switch in the high-speed position, the resistors are bypassed and the motor will operate normally, but only in high speed.

Answer D is wrong. With the resistors removed, current cannot flow from the low-speed switch position to the motor.

Question #100

Answer A is wrong. If the red wire was open a 24-volt reading would not be present at the ballast.

Answer B is wrong. The original lamp was replaced with a known good bulb and still there was no light.

Answer C is wrong. If the black wire was open a 24-volt reading would not be present at the ballast.

Answer D is correct. There is power and ground at the lamp ballast, which indicates that the circuitry up to and after the lamp ballast is good. A continuity check on the green wire proved that it was good, and both sockets were found in tact. A known good lamp was also installed. By process of elimination the lamp ballast itself is all that is left in the circuit that could most likely cause the problem.

Question #101

Answer A is wrong. This would test the voltage drop in the positive battery cable.

Answer B is wrong. This would test the combined voltage drop of both the positive battery cable and the solenoid internal contacts.

Answer C is correct. By probing between points B and M, the voltage drop across the solenoid internal contacts can be tested with the starter cranking.

Answer D is wrong. Probing between points G and ground will only test the voltage drop in the solenoid ground wire.

Question #102

Answer A is wrong. Swapping the gauge (or any suspected defective part) with a new one is not a good practice without first determining the fault.

Answer B is wrong. As with the previous answer, it is bad practice to troubleshoot a problem by trial and error.

Answer C is wrong. With many senders on an electronic engine, grounding is not possible.

Answer D is correct. By using the diagnostic tool, you can troubleshoot the problem using the OEM-recommended procedure. If the diagnostic tool shows normal readings, then the gauge would be suspect.

Question #103

Answer A is wrong. A starting switch is considered part of the control circuit.

Answer B is correct. The starter motor is not considered part of the control circuit. It is the device being controlled.

Answer C is wrong. The starting-safety switch is part of the control circuit. If it is not closed, the circuit is open.

Answer D is wrong. The magnetic switch is also part of the control circuit. Power from the control circuit is fed to the magnetic switch to activate the starter solenoid.

Question #104

Answer A is wrong. An open in the insulated side of the circuit should be checked upstream (toward the source) from an inoperative component.

Answer B is wrong. An ammeter is used to check current flow. It is impractical to use it as a method of checking continuity.

Answer C is wrong. Both Technicians A and B are wrong.

Answer D is correct. Both Technicians A and B are wrong.

Question #105

Answer A is correct. This picture shows a voltmeter measuring the output voltage of the charging system.

Answer B is wrong. To perform a positive charging cable voltage drop test, one voltmeter probe would be at the "battery" terminal on the starter solenoid, and the other probe would be at the output terminal of the alternator.

Answer C is wrong. To measure the voltage drop of the charging ground circuit, you would probe at the alternator housing with one lead, and at the battery negative terminal with the other lead.

Answer D is wrong. To measure starter operating voltage, you would probe at the starter, not the alternator.

Question #106

Answer A is wrong. If the ground side of the motor shorted to ground, there would be no adverse reaction. This system does not use ground side switching.

Answer B is wrong. If the control switch shorted to ground, the fuse would blow.

Answer C is correct. If the switch contacts are stuck closed, there would be a continuous circuit and constant pump operation.

Answer D is wrong. There is no separate relay in this system for the wiper-washer pump motor.

Question #107

Answer A is wrong. All 12-volt systems used on medium-duty buses use DC current for motor power.
Answer B is wrong. Installing a relay between the blower motor resistors and the motor would defeat the purpose of the resistors, which is to decrease the speed of the motor.
Answer C is wrong. Neither technician is right.
Answer D is correct. Both technicians are wrong.

Question #108

Answer A is wrong. A battery that is being load tested must have a reading above 9.6 volts under load, not less, in order for it to pass the test.
Answer B is wrong. A battery should be load tested to three times its ampere-hour rating, not two times.
Answer C is wrong. Neither technician is correct.
Answer D is correct. Both technicians are wrong.

Question #109

Answer A is wrong. While theoretically the resistance of the wire could be measured with an ohmmeter, a voltage drop test is a far more accurate way of assessing the condition of this wire.
Answer B is wrong. An ammeter will only show the amount of current flowing through the circuit; it will not give a good indication of the resistance of the particular wire. Excessive resistance could be in the motor itself, but the ammeter would not help to isolate this.
Answer C is correct. A voltage drop test with the circuit under load is the best way to determine high resistance.
Answer D is wrong. Simply measuring the voltage at the end of the wire would not isolate the problem to the wire itself. Low voltage could be due to a weak battery, but you could not detect this unless you measured voltage at either end of the wire simultaneously.

Question #110

Answer A is wrong. There is no voltage solenoid in a bus charging system.
Answer B is correct. A voltage regulator is an integral part of the charging system, whether it is external or internally mounted in the alternator.
Answer C is wrong. There is no voltage transducer in a bus charging system.
Answer D is wrong. A magnetic switch is part of the cranking circuit, not the charging circuit.

Question #111

Answer A is wrong. This test would be measuring the voltage drop in the regulator ground circuit.
Answer B is correct. This outlines the proper test procedure for measuring voltage drop on the ground side of the charging circuit. The ground side of the circuit should also be tested. Move the voltmeter leads to the alternator casing and the battery ground pole. The voltage drop in this case should be 0. If not, the alternator mounting might be loose or corrosion might be built up between the casing and mounting bracket.
Answer C is wrong. Only Technician B is right.
Answer D is wrong. One of the technicians is correct.

Question #112

Answer A is correct. The magnetic pickup shown in the picture is used for shaft speed sensing, such as speedometers. The sensor contains a permanent magnet that creates a magnetic field that when broken by a rotating gear generates an AC voltage. The gauge assembly or computer module counts the impulses and then electronically computes the vehicle speed in miles per hour and the engine speed in revolutions per minute, or rpm. This information is used to drive the speedometer gauge to indicate the specific readings. Most of these speedometers use the electromagnetic air core type of gauges.
Answer B is wrong. The figure does not show a pressure sensor.
Answer C is wrong. The figure does not show a temperature sensor.
Answer D is wrong. The figure does not show a level sensor.

Question #113
Answer A is wrong. If the fuse keeps blowing, this means that there must be a short somewhere to ground, or a faulty bulb. Checking between points A and B will not indicate a short to ground.
Answer B is correct. If a short to ground is suspected, then the ohmmeter must be placed at some point in the circuit with respect to ground. An unusually low reading will indicate a short to ground problem.
Answer C is wrong. Only one of the technicians is correct.
Answer D is wrong. One technician is correct.

Question #114
Answer A is wrong. A short to ground on the ground side of the bulb will not change overall circuit resistance. Therefore, current flow will not increase.
Answer B is correct. A short to ground on the ground side of this bulb has effectively taken the switch out of the circuit. This particular circuit uses ground-side switching, but in this case, the short has bypassed the switch.
Answer C is wrong. Only Technician B is correct.
Answer D is wrong. One of the technicians is right.

Question #115
Answer A is wrong. If the solenoid were not activated, the starter motor would not spin.
Answer B is correct. A defective drive may not crank the engine even though the starter motor spins.
Answer C is wrong. Only Technician B is correct.
Answer D is wrong. One of the technicians is correct.

Question #116
Answer A is correct. A 2-amp draw through a battery with the engine off will cause a battery to discharge overnight.
Answer B is wrong. Most electronic control modules draw less than 50 milliamps with the master/key switch off.
Answer C is wrong. Only Technician A is correct.
Answer D is wrong. One of the technicians is correct.

Question #117
Answer A is wrong. Brake interlock connections are not part of the starter control circuit.
Answer B is wrong. The starter motor is not considered part of the control circuit. It is another high-amperage draw component.
Answer C is wrong. The starter solenoid high-current switch contacts are not part of the control circuit; rather they switch the high-current circuit from the battery to the starter motor itself.
Answer D is correct. The magnetic switch is part of the starter control circuit. This includes anything between the starting switch and the starter solenoid, not including the switching contacts.

Question #118
Answer A is wrong. A faulty switch would have to affect the taillights on both sides equally.
Answer B is correct. A poor ground connection at the left-rear taillight may cause excessive resistance and low-current flow, therefore causing a dim light.
Answer C is wrong. Only Technician B is correct.
Answer D is wrong. One of the technicians is correct.

Question #119
Answer A is wrong. Older-type 2-speed wiper motors use high and low speed brushes. However, Technician B is also correct, so answer A is wrong.
Answer B is wrong. Older-type 2-speed wiper motors use an external resistor pack to control the speeds. However, Technician A is also correct, so answer B is wrong.
Answer C is correct. Both technicians are correct.
Answer D is wrong. Neither technician is wrong.

Question #120

Answer A is wrong. The compressor clutch can only turn in one direction regardless of the application of the clutch. The driving engine cannot turn backward.

Answer B is wrong. It is the A/C circuit that is being protected, not the clutch itself. The voltage spike is redirected back into the clutch during disengagement.

Answer C is correct. The purpose of the diode is to protect the rest of the circuit from a voltage spike generated when the clutch magnetic field collapses.

Answer D is wrong. The diode cannot limit current flow to the clutch because a diode is simply a one-way check valve, not a resistor. Even if the diode acted as a resistor, it would have to be wired in series with the clutch, not in parallel, to affect the current flow.

Glossary

Actuator A device that delivers motion in response to an electrical signal.

AH Ampere-Hours, an older method of determining a battery's capacity.

Alternator A device that converts mechanical energy from the engine to electrical energy used to charge the battery and power various vehicle accessories.

Ammeter A device (usually part of a DMM) that is used to measure current flow in units known as amps or milliamps.

Ampere A unit for measuring electrical current, also known as an amp.

Analog Signal A voltage signal that varies within a given range from high to low, including all points in between.

Analog-to-Digital Converter (A/D converter) A device that converts analog voltage signals to a digital format, located in the section of a control module called the input signal conditioner.

Analog Volt/Ohmmeter (AVOM) A test meter used for checking voltage and resistance. These are older style meters that use a needle to indicate the values being read. Should not be used with electronic circuits.

Armature The rotating component of a (1) starter or other motor, (2) generator, (3) compressor clutch.

Diagnostic Reader The diagnostic tool supplied by most manufacturers for accessing data information from various electronic systems in buses.

Blade Fuse A type of fuse having two flat male lugs for insertion into mating female sockets.

Blower Fan A fan that pushes air through a ventilation, heater, or air conditioning system.

Cartridge Fuse A type of fuse having a strip of low melting point metal enclosed in a glass tube.

CCA Cold cranking amps, a common method used to specify battery capacity.

CCM Chassis control module, a computer used to control various aspects of driveline operation, usually does not include any engine controls.

Circuit A complete path for an electrical current to flow.

Circuit Breaker A circuit protection device used to open a circuit when current in excess of its rated capacity flows through a circuit. Designed to reset, either manually or automatically.

Data Bus Data backbone of the chassis electronic system using hardware and communications protocols consistent with CAN 2.0 and SAE J-1939 standards.

Data Link A dedicated wiring circuit in the system of a vehicle used to transfer information from one or more electronic systems to a diagnostic tool, or from one module to another.

Diode An electrical one-way check valve. It allows current flow in one direction but not the other.

DMM Digital multimeter, a tool used for measuring circuit values such as voltage, current flow, and resistance. The meter has a digital readout, and is recommended for measuring sensitive electronic circuits.

ECM/ECU Electronic control module/unit, acronyms for the modules that control the electronic systems on a bus.

Electricity The flow of electrons through various circuits, usually controlled by manual switches and senders.

Electronically Erasable Programmable Memory (EEPROM) Computer memory that enables write-to-self, logging of failure codes and strategies, and customer/proprietary data programming.

Electronics The branch of electricity where electrical circuits are monitored and controlled by a computer, the purpose of which is to allow for more efficient operation of those systems.

Electrons Negatively charged particles orbiting every atomic nucleus.

EMI Electromagnetic interference, low level radiation that interferes with electrical/electronically-controlled circuits, causing erratic outcomes.

Fault Code A code stored in computer memory to be retrieved by a technician using a diagnostic tool.

Fuse A circuit protection device designed to open a circuit when amperage that exceeds its rating flows through a circuit.

Grounded Circuit A condition that causes current to return to the battery before reaching of its intended destination. Because the resistance is usually much lower than normal, excess current flows and damage to wiring or other components usually results. Also known as short circuit.

Halogen Light A lamp having a small quartz/glass bulb that contains a filament surrounded by halogen gas. It is contained within a larger metal reflector and lens element.

Harness and Harness Connectors The routing of wires along with termination points to allow for vehicle electrical operation.

High-Resistance Circuits Circuits that have resistance in excess of what was intended. Causes a decrease in current flow along with dimmer lights and slower motors.

In-line Fuse A fuse usually mounted in a special holder inserted somewhere into a circuit, usually near a power source.

Insulator A material, such as rubber or glass, that offers high resistance to the flow of electricity.

Integrated Circuit A solid state component containing diodes, transistors, resistors, capacitors, and other electronic components mounted on a single piece of material and capable of performing numerous functions.

Jumper Wire A piece of test wire, usually with alligator clips on each end, meant to bypass sections of a circuit for testing and troubleshooting purposes.

Jump Start A term used to describe the procedure where a booster battery is used to help start a vehicle with a low or dead battery.

Magnetic Switch The term usually used to describe a relay that switches power from the battery to a starter solenoid. It is controlled by the start switch.

Maintenance Free Battery A battery that does not require the addition of water during its normal service life.

Milliamp 1/1000th of an amp. 1000 milliamps = 1 amp.

Millivolt 1/1000th of a volt. 1000 millivolts = 1 volt.

Ohm A unit of electrical resistance.

Ohmmeter An instrument used to measure resistance in an electrical circuit, usually part of a DMM.

Ohm's Law A basic law of electricity stating that in any electrical circuit, voltage, amperage, and resistance work together in a mathematical relationship.

Open Circuit A circuit in which current has ceased to flow because of either an accidental breakage (such as a broken wire) or an intentional breakage (such as opening a switch).

Output Driver An electronic on/off switch that a computer uses to drive higher amperage outputs, such as injector solenoids.

Parallel Circuit An electrical circuit that provides two or more paths for the current to flow. Each path has separate resistances (or loads) and operates independently from the other parallel paths. In a parallel circuit, amperage can flow through more than one load path at a time.

Power A measure of work being done. In electrical systems, this is measured in watts, which is simply amps · volts.

Processor The brain of the processing cycle in a computer or module. Performs data fetch-and-carry, data organization, logic, and arithmetic computation.

Programmable Read Only Memory (PROM) An electronic memory component that contains program information specific to chassis application: used to qualify ROM data.

Random Access Memory (RAM) The memory used during computer operation to store temporary information. The computer can write, read, and erase information from RAM in any order, which is why it is called random. RAM is electronically retained and therefore volatile.

Read Only Memory (ROM) A type of memory used in computers to store information permanently.

Reference Voltage The voltage supplied to various sensors by the computer, which acts as a baseline voltage; modified by sensors to act as an input signal.

Relay An electrical switch that uses a small current to control a large one, such as a magnetic switch used in starter motor cranking circuits.

Reserve Capacity Rating The measurement of the ability of a battery to sustain a minimum vehicle electrical load in the event of a charging system failure.

Resistance The opposition to current flow in an electrical circuit; measured in units known as ohms.

Rotor (1) A part of the alternator that provides the magnetic fields necessary to generate a current flow. (2) The rotating member of an assembly.

Semiconductor A solid-state device that can function as either a conductor or an insulator depending on how its crystalline structure is arranged.

Sensing Voltage A reference voltage put out by the alternator that allows the regulator to sense and adjust charging system output voltage.

Sensor An electrical unit used to monitor conditions in a specific circuit to report back to either a computer or a light, solenoid, etc.

Series Circuit A circuit that consists of one or more resistances connected to a voltage source so there is only one path for electrons to flow.

Series/Parallel Circuit A circuit designed so that both series and parallel combinations exist within the same circuit.

Short Circuit A condition, most often undesirable, between one circuit relative to ground, or one circuit relative to another, connect. Commonly caused by two wires rubbing together and exposing bare wires. It almost always causes blown fuses and/or undesirable actions.

Signal Generators Electromagnetic devices used to count pulses produced by a reluctor or chopper wheel (such as teeth on a transmission output shaft gear) which are then translated by an ECM or gauge to display speed, rpm, etc.

Slip Rings and Brushes Components of an alternator that conduct current to the rotating rotor. Most alternators have two slip rings mounted directly on the rotor shaft; they are insulated from the shaft and each other. A spring-loaded carbon brush is located on each slip ring to carry the current to and from the rotor windings.

Solenoid An electromagnet used to perform mechanical work, made with one or two coil windings wound around an iron tube. A good example is a starter solenoid, which shifts the starter drive pinion into mesh with the flywheel ring gear.

Starter (Neutral) Safety Switch A switch used to insure that a starter is not engaged when the transmission is in gear.

Switch A device used to control current flow in a circuit. It can be either manually operated or controlled by another source, such as a computer.

Transistor An electronic device that acts as a switching mechanism.

Volt A unit of electrical force, or pressure.

Voltage Drop The amount of voltage lost in any particular circuit due to excessive resistance in one or more wires, conductors, etc., either leading up to or exiting from a load (e.g., starter motor). Voltage drops can only be checked with the circuit energized.

Voltmeter A device (usually incorporated into a DMM) used to measure voltage.

Watt A unit of electrical power, calculated by multiplying volts · amps.

Windings (1) The three separate bundles in which wires are grouped in an alternator stator. (2) The coil of wire found in a relay or other similar device. (3) That part of an electrical clutch that provides a magnetic field.

Xenon Headlights High voltage, high-intensity headlamps that use heavy xenon gas elements.

Notes

Notes

Notes

Notes

Notes

Notes

Notes

Notes

Notes

Notes

Notes

Notes